75

LEFT AND RIGHT IN SCIENCE AND LIFE

VILMA FRITSCH

LEFT AND RIGHT IN SCIENCE AND LIFE

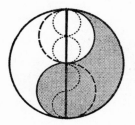

BARRIE & ROCKLIFF

LONDON

S.B.N. 214 65915 1

LINKS UND RECHTS
IN WISSENSCHAFT UND LEBEN
© 1964 by W. Kohlhammer GmbH, Stuttgart
English translation
© 1968 by Barrie Books Ltd
First published 1968 by
Barrie & Rockcliff
2 Clement's Inn, London WC2
Printed in Great Britain by
Western Printing Services Ltd, Bristol

CONTENTS

PREFACE

Motto: "If right and left had not been relegated to the traffic regulations
or to the terrestrial and celestial ceremonial, Science and
Philosophy would have known how to use them fittingly."
(Silvio Ceccato)

In 1957 the two Chinese scientists, Lee and Yang, received the
Nobel Prize for Physics for proving the non-conservation of
parity. For a few days the newspapers talked of left and right,
for the law of conservation of parity had laid it down, so the
physicists explained, "that in nature there was no preferential
treatment of the right/left screw sense". Many of us then began
to wonder about the meaning of right and left. What actually do
we mean when we speak of right and left? Have these concepts
any meaning at all? They can hardly be completely meaningless.
In daily life we orient ourselves towards right and left. Biologists
and doctors speak of right- and left-handed people, and the
mathematician speaks of the right and left side of an equation.
There is no science that does not use these terms. Even in
politics one talks of right and left.

An answer to our questions has already been given by Hermann Weyl in his Princeton Lectures, which were published
under the title *Symmetry*.[1] In an inimitable, clear and most impressive fashion he proclaims the triumph of symmetry. However, death prevented the great German mathematician from
dealing with the epistemological consequences of this new discovery, which many regarded as implying the collapse of the
bilateral symmetry of space. It is to be hoped that someone else,
who like Weyl has helped to build up the science of nuclear

7

physics, will one day write the missing chapter. I, for my part, will admit that this "missing chapter", or in other words the need to clarify a situation created by the attack of physics on certain basic concepts of knowledge, has led me to write this book.

Is it not strange that the discussion concerning the meaning of right and left has not been settled long ago? For centuries philosophers and men of science have been trying to demystify both these concepts and to show that they have, in actual truth, no "meaning" at all but are simply schemes of order and nothing else. Nevertheless, there is something within us that will not accept this view. Whenever we think we have grasped the fact that Nature, though she may perhaps distinguish between right and left, shows no actual preference for either of them—whenever this happens, I say, we are conscious of a certain uneasiness, of a doubt. To remove that doubt, both from my own mind and that of others, I should like here to sketch out a survey of the problems which right and left entail. This survey will seek to determine the possible meanings of right and left in other fields than that of physics and upon other planes than the physical.

In Hermann Weyl's important work there are numerous examples of symmetry in science and art, and, above all, in the art forms of Nature. But Weyl's aim is to pass, by way of the precise concept of geometrical symmetry, from the συμμετρία of the Greeks—understood by them as harmony—to that conception of symmetry which must ultimately govern the whole of our thinking. This prevents him from dwelling very long on the less mathematical and less "symmetrical" sciences. It is precisely to these last that this book is dedicated. It therefore stands not so much under the sign of symmetry as under that of a certain asymetry. For if the physicists were reluctant to abandon the idea of a "difference in the regions of space"—to quote Kant—biologists and psychologists were even more hesitant, since their conception of left and right is situated somewhere between the absolute symmetry of logical mathematical relationships and the emotionally charged polarity of pre-scientific symbolism.

We must not expect a solution, let alone a number of solutions, but this does not mean that our survey has no more than a curiosity value. Right and left, appearing at all levels of organization, could prove to be key concepts. Certainly they will lead us everywhere to the problems at the ultimate frontier of things, to the problems of origin, in short to the great problems which, apart from the fact that they have remained unsolved, have this in common: that first of all they need to be posed correctly.

Let me just add this remark. The greater and more profound the problem, the more people will believe in their right to discuss it. The present volume should make it plain that, although the ultimate syntheses may be philosophical ones, it will always pay us to start by letting the natural sciences say what they have to say. This alone, I hope, will justify this survey of the right/left problems, which is designed to inform the layman—and in some matters we are all of us laymen—about a few well known, or possibly less well known, results of research in different fields of science.

I

THE PAIR

1. *Forms of duality*

Right and left form a pair. This brief formula clearly exhausts our knowledge of right and left, though we must of course understand "pair" in its proper sense and open the proper door with the key of this word. For there are many pairs and they are to be found at the most widely different levels; light and darkness; heaven and earth; spirit and matter; sensuality and reason; right and wrong; man and woman; rest and movement; space and time; subject and object; form and function; good and evil; warmth and cold; black and white. Some of these contrasts, such as warmth and cold, have been exposed by science as being only apparently opposites. Of others we believe, rightly or wrongly, that they have some kind of an existential status "as a pair".

There can be no doubt whatever that the most profound problems of modern thought centre on the concept of duality, of the pair. It is often said—and in the last century it was said more often still—that philosophy and science strive towards unity. In reality however things are not "all one" either to the philosopher or to the scientist. The physicist, for instance, seeks to simplify and unify, but he does not by any means seek for a final identity, and when his colleague in the science-fiction story, shortly

before the world catastrophe, brings the one and only world
formula to safety, he can only smile at him; he knows that this
"super-Einstein" is making a completely useless present to the
future inhabitants of the earth. The new ideal should indeed be
called "complexity" rather than unity. Gaston Bachelard,[1] who
penetrated as did few others into the inmost "Spirit of the New
Science", shows us by the example of optics, of mechanics and
microphysics, how modern non-Cartesian thought no longer
endeavours to "explain" the world but instead complicates and
enriches experience and discovers multiplicity, and more
especially duality, behind an illusory simplicity conjured up by
"compensated problems". Indeed he seeks to introduce as an
espistemological principle the idea that "complementary
attributes should be included in the nature of being and that we
should do away with the tacit assumption that being is always a
sign of unity".[2]

Philosophy and psychology already seem to a large extent to
meet this demand. In various forms and shades of meaning this
question is being asked: which can claim ontological or psychological priority: the individual or the pair, the subject or the
object? Or can both claim it equally? Can it be claimed by the
"self" or by the "other" or by an indivisible duality of self and
other? And, strangely enough, the answer supplied by not a few
to this question (which, incidentally, is much more profound
than the well known scholastic problem, which came first the
hen or the egg?) is: *"in the beginning was the pair"*.

All recent "dual philosophies"—it is best to avoid calling them
"dualistic"—have this in common: they tend to exalt pairhood,
even to treat it with a kind of emotionally charged veneration.
"When I see a cell dividing itself", says the geneticist, William
Bateson, "I experience much the same feeling as an astronomer
who has witnessed the birth of a double star. An *aboriginal act of
nature* is there taking place before my eyes." True, a biologist is
here speaking of a biological phenomenon, but this feeling of
cosmic reverence also characterizes the affirmation of duality
among the philosophers of our time. The dialectical philosopher

like his existentialist colleague feels no regret when contemplat-
ing the fragmented nature of the world. These men do not yearn
for unity. Unity is for them immobility, ending and death. But it
is precisely *the fact of being two and not one* that guarantees progress
to the dialectician, releases the world for him from its rigidity
and makes changes possible, while it gives to the existentialist
that modest form of freedom which derives from the acceptance
of the basic paradox of human existence and from standing
firm in the face of the irreconcilable dualism of our life situation.

By seeing the relationship between "self" and "other" as the
"constitutive condition" of consciousness, Hegel and Husserl
endowed duality with an *ontological dignity* which since then it has
never lost. According to Saitre the "other" is necessary because
one can only define oneself with reference to another. Perhaps
"the look given me by the other" will, as in Sartre's case, only
reveal me to myself in my "state of shame"; nevertheless it alone
can and does reveal this self of mine to me, and that is what
matters.

Though coming from a different starting point, travelling
a different road and pursuing a different goal from that of
existentialist analysis, psychological analysis also leads to a
revaluation of duality. Contemporary psychologists agree that
psychology must be written not only in the singular and plural,
but in the dual, a form that still survives in some languages. The
dialectic between "self" and "other" provides them with a
valuable model for research into the empirical personality of
man, a personality built up through a perpetual confrontation
of self and other. *"Duality is anterior to unity,"* says Henri Wallon,
for instance.[3] His pupil, Rene Zazzo,[4] used the natural experiment provided by the existence of twins with its often contrasting
pair effects as a means of illuminating psychological individuality and of showing that the formation of personality does not
depend exclusively on the factors of heredity and environment.

More directly connected with the questions dealt with here
are various contemporary enquiries into the nature and genesis
of thought. For Wallon and his pupils there is no thought that

can be formulated beyond the pair and its dialectic. The principle of all thought resides in a *binary structure*, in the basic relationship represented by the pair. According to the French psychologist "thinking in pairs" is by no means only an archaic form of thought to be found at a certain stage of childish development. It is also alive in the thinking of adults even in its highest scientific constructions.

It has been contended that the significance of pairhood was first perceived by the Neo-Hegelian, Ludwig Feuerbach, and that it was also reflected in the romantic outpourings of Coleridge. But before Feuerbach, Schelling had discovered the elements of his dual theogeny and philosophy in Greek mythology and in the third chapter of Genesis and, following up Jacob Boehme's basic idea of "Selbstentzweiung", self-division, had recognized "a true duality" in God. Going back step by step one could trace a very long "history of duality". Strictly speaking it goes back to the earliest mythical times, to that "time that was before time existed". After all, before man engaged in philosophy he intuitively expressed in myths abstract relationships and even, according to some authorities, concrete biological realities.[5] Equally full of insights were his "sacred mathematics" (which served quite different purposes from those of measurement and calculation), and the unspoken myth of his ritual.

We encounter the rudimentary category of the pair in the myths of most peoples: divine progenitors of a tribe, city-founding twins, hermaphrodites, androgynes, doubles, shadows and mirror-images, all bear witness to the duality of the world. And the modern poet, seeking to express the numinous, still reaches back towards these archetypal figures and seeks to discover their message.

The *myth of the founder couple* is, according to Zazzo, an "archaic dream" in which the twinship theme and the idea of hermaphroditism are fairly closely intermingled, "as though the unity of man could only be understood through the united character of the pair, and, contrariwise, the oneness of the pair could only be explained through the essential unity of the two partners". The mythology of *Egypt* is particularly rich in such aboriginal pairs, whether brother and sister or husband and wife. Isis and Osiris are merely the best known example. There are actually quite a number of them, four going back to before the creation: a god and a goddess for each of the properties of chaos. Attempts have been made to connect this with the symmetrical lay-out of the country, the right and left banks of the Nile being mirror-images of one another. In ancient China the mythical first ruler and soothsayer, Fu-Hsi, is the husband and brother of Nü-Kua and it is to this aboriginal pair with its dual character of unity and differentiation that the invention of marriage is ascribed. It was only the divine hierogamy that made human marriage possible. Pictorial representations show us Fu-Hsi and Nü-Kua with the lower parts of their bodies intertwined; Fu-Hsi holds the T-square in his right hand, Nü-Kua the compasses in her left.

Even in what are definitely myths and legends about twins two themes occur, either separately or combined, the theme of the similarity of identical twins and the theme of twins of different sex—or the theme of the *double* (the search for one's own self) and the theme of the *pair* (union with the other self). Zazzo, who explores both these themes, characteristically holds that the second is the older "... as though the consciousness of the bisexual pair as an original unity preceded that of individuality—which last is symbolized by the double".

Plato's beautiful myth in the Symposium is perhaps an artificial one, but it uses genuinely mythical ideas and is much more than a mere allegory. It tells how in ancient times we humans had two pairs of arms and legs and were completely "double" like Siamese twins. By this mythical tale Plato seeks to throw light on the aboriginal unity of the sexes, a unity that preceded sexual differentiation. At the same time he gives us— in addition to a defence of homosexuality—a theory of the origin of love which is still valid today. As we shall see, this myth is also a right-left myth, for it illustrates, according to Herman Weyl, the degeneration of rotation symmetry into the more specialized bilateral symmetry. Robert Merle[6] holds that the Bible too contains such a myth of the transition from spherical to bilateral symmetry. Merle declares that the word translated as "rib" in the Creation story has only to be rendered as "side", as is actually done by Ibn Seoud and Ibn Fataquera in their commentaries on Maimonides, "for us to rediscover in Adam a double being, an androgynous creature similar to those referred to in the Symposium or those found in the Babylonian and Phoenician Cosmogenies".

This aboriginal duality, which we find in the myths of twins and in those which tell of pairs of tribal ancestors, occurs in more abstract form in the sacred mathematics of ancient peoples. The qualitative mathematics of the Pythagorean tradition which subsisted alongside of ordinary quantitative mathematics, recognized values, both positive and negative, in numbers, depending upon whether the numbers were odd or even. The odd numbers "corresponded" to the limited, the one, the male, light and motion, the straight as distinct from the crooked, the good and the right. The even numbers "corresponded" to the unlimited, the many, the female, darkness, evil and *the left*. We know relatively little about this number-mysticism, but it is probably correctly considered to be the prototype of all those "Manichaean" views of the world in which the world is divided up between two mutually hostile principles, an absolute good and an absolute evil.

We should in no wise overlook the fact that there are other divisions into two categories, divisions which do *not* possess this particular character. Thus we encounter in ancient China a similar ordering of numbers by pairs, a system which departs from our own "natural" sequence of numbers, in which each number comes into being out of the preceding one through the addition of unity. These holy numbers no more serve the purposes of calculation than do the Pythagorean ones, though both the ancient Chinese and the Pythagoreans could, of course, calculate very well indeed. But both good and evil are absent in the list of pairs. Recent research has also shown that they really do not form a rigid framework at all but constitute a sort of dynamic pattern marked by the rhythmical exchange of qualities that beget each other. Here odd and even are not complete opposites, and the even contains the odd and is able to bring it forth. The dynamic character of ancient Chinese number symbolism is perhaps seen in the most striking fashion in an instrument used by the diviners. This consists of two little plates of hard and soft wood which can be rotated one against the other. Upon these there are all manner of symbols arranged like the figures on a magic square. In this little mechanical toy the strict symmetrical scheme of the Pythagoreans is literally set in motion. The Yin-Yang symbol illustrated on the title page admittedly does not move but it is not hard to imagine it turning; it calls up the same associations as the Mandala[7]—the symbols of the centre and of the union of opposites which are correlated with the number 4, or double duality.

The Yin Yang doctrine of the ancient Chinese deserves our very special attention, for it is the most "modern" of all the immemorial dualisms. The words Yin Yang have no exact equivalent in any European language. What exactly is it that they express? The Jesuits who in the seventeenth century were the first to discover the world of Chinese thought had of course seen it through the spectacles of European Christian substantialism and turned Yin and Yang into "principles". But according to Marcel Granet's authoritative work[8] Yin and Yang are neither principles nor substances nor powers, nor are they any other kind of Scholastic entity. Rather are they emblems whose social origin and concrete value always remain discernible. In ancient times, when the corporations of plowmen and women weavers met annually at the time of the vernal equinox for dancing, alternating songs, competitive games and orgiastic union, Yang meant the sunny river bank, Yin the one that lay in shadow, Yang meant life in the fields in summertime, Yin the darkness and security of the winter refuge. By means of their direct suggestive power these two signs regulated the drama of these two complementary groups who were separated by sex,

vocation, place and time, and at the same time the drama of the
universe: Now Yin, now Yang! Granet thus considers the
concepts Yin and Yang as representing a totality of images
governed by the idea of rhythm. They call up the idea of a
practical, concrete classification and distribution which makes
it possible to order the world as though it were a play, to adapt
oneself to that world and keep in harmony with it: a world
which, says Granet, "shows no phenomenon that does not
correspond with a cyclic totality formed by the conjunction of
two alternating and complementary manifestations". Thus Yin
and Yang are not complete opposites. "Nothing", Granet
concludes "would be more contrary to the spirit of ancient
China—and up to a point of modern China too—than to set up
Yin and Yang against each other as we would set up 'being'
and 'not being' or sacred and profane. *Above all the Chinese are
not subject to that religious zeal which condemns us to divide everything up
between good and evil.*"

We thus encounter a philosophy among the ancient Chinese
that is very different from the Manichaeanism that can be
discerned in the Pythagorean number mysticism and indeed
very different from Manichaeanism as such, using that term in
the widest sense as expressing a basic human attitude that we are
liable to meet at any place and at any time . . . Joseph Needham
in the second volume of his great work *Science and Civilization in
China*[9] lays stress on the fact that in the Ying-Yang doctrine there
is no trace of any undertone of good and evil. It was precisely
this that made it an effective instrument for the maintenance of a
true balance and for the achievement of such valid human aims
as happiness, health and good order. According to Needham's
bold thesis, such non-Manichaean thinking is also destined to
play an important part in science—and it has indeed already
begun to do so, thanks to the mediatory activity of Leibniz. In
his view the ancient Chinese manner of forming symbols
displays some astonishing analogies with the very latest thought.
It is true that the Chinese proved incapable of producing our
classical science. Yet they shot many an arrow close to the spot

on which Niels Bohr, Max Planck and Albert Einstein were one day to stand.

We can find support and confirmation for the views of Granet and Needham in the studies made by Werner Müller[10] on the opposite side of the world from China, among the Indians of North America. In his book *Die Religionen der Waldlandindianer Nordamerikas* (The Religions of the Forest Indians of North America) he opens up a second "anti-Manichaean" cultural region, the true nature of which has remained concealed from us till quite recent times—and for the same reason that caused this to happen with regard to China, for here too the observers projected their own cast of mind onto the subject of their study or "succumbed to their own compulsive ideas". The Indians know of no Devil or personification of evil. Their "philosophy of the pair" is very Chinese and so is their conception of life as a harmonious co-operation of cosmic beings. Doubleness is an essential trait of the Janus-like figures in their traditions just as it is in the ancient Babylonian *Ea*, the Greek *Hermes* and the Germanic *Loki*. Werner Müller points out that "we encounter here a religious philosophy which is closely bound up with the economic side of tribal life, with the division of labour between the sexes in hunting and agriculture, with the dual aspects of material existence" (ibid. p. 67).

2. *Symmetry and polarity*

These necessarily fragmentary introductory remarks show how numerous and how subtly different are the various pair relationships, oscillating between the two extremes of complete similarity and diametrical opposition; they also show the intimate connection of such pair relationships with the roots of social and religious life.

It is no easy task with such a wealth of examples of philosophic

and mythical thinking to pick out the particular model to which
the pair *right-left* corresponds. Are right and left qualitatively
different as were man and woman in the original pair? If we go
still further, is one good, the other bad? Are they like twins,
similar in every detail yet distinct individuals? Can they be
separated or are they inseparable in the sense of the Chinese
proverb "When the lips are missing the teeth freeze"? Do they
constitute together two aspects of the same dynamic order
(Yin and Yang)? Are right and left after all nothing but words,
symbols, free creations of the human mind, labels which—
with or without good reason—we stick on things?

As we endeavour to bring order into the great multitude of
suggested examples two concepts force themselves upon our
minds: *symmetry* and *polarity*. These two concepts seem at first
sight to be singularly well suited to express the different shades of
pairhood. For most of us they have a familiar ring. When we say
"Right and Left are *polar opposites*" we are obviously trying to
ascribe to left and right a qualitative difference, we are trying to
regard them as distinct "substances"—using that term in its
philosophical sense. If we recognize the symmetry of right and
left we deny them the quality of substances or at any rate we
accord their intrinsic nature an inferior place and direct our
gaze to the *relation* between them. On closer examination
however symmetry and polarity prove to be categories which
can only with difficulty be compared with one another—and
perhaps cannot be compared at all; they "talk past each other".
Being neither polar opposites nor symmetrical, they themselves
do not form a pair. It is significant that polarity and symmetry
have suffered completely different fates.

Symmetry has gone from strength to strength, as Weyl shows
in his book. Starting out as a nebulous aesthetic notion, something
that was wholly relegated to the realm of "feeling", it became a
precise concept of mathematics, the most important among the
natural sciences of today.

Originally the Greek συμμετρία was nothing more than what
we call proportion, harmony, balance, and was by no means

confined to spatial objects. As Weyl points out, the synonym "harmony" points more towards an acoustic or musical application. Being still close to the immediacy of feeling, symmetry demanded only the approximations of *resemblance*. Perhaps the etymologist could tell us when the requirement of resemblance was replaced by the stricter one of "similarity", and whether the transition of symmetry from the realm of "pleasant sounds" and of "formal beauty to the realm of gods and ideas, and of mathematical-logical abstractions was sudden or gradual. However that may be, in defining symmetry today we use the term "similarity". Thus in an issue of the *Studium Generale*[11] which is almost entirely devoted to the subject of symmetry, W. von Engelhardt describes symmetry as "the spatial and/or temporal repetition of similar elements".

This definition is in line with the use of the concept of symmetry in classical science. The layman is puzzled by this, since, strangely enough, the concept has for him become entirely spatial, and symmetry in time, which perhaps was its more original form, seems to him not to be symmetry at all. Moreover, the person who is himself more or less right-left-symmetrically endowed, naturally regards right-left symmetry as the preferred, or even the only kind of symmetry. To those of us who are not mathematicians, the word symmetry is more likely to conjure up the picture of a mantelpiece graced by two candlesticks than that of a sea-urchin or of Einstein's theory of relativity. And yet both the sea-urchin and Einstein's theory can claim to be called symmetrical. It is widely held that we can perceive symmetry only between right and left or that we can perceive this better than that between above and below. Pascal attributes this to the anthropomorphic origin of the concept—as does also Mach, for whom this "fact" seems of very great importance. Many people however, are quite unaware of any difference in this matter and the most perfect symmetry is expressed for them by a group of trees mirrored in a clear mountain lake.

Interesting as such reflections may be to the student of aesthetics or psychology, they have, I need hardly point out, nothing to do with the mathematical and geometrical definition of symmetry; for this last emancipates itself from all anthropomorphic images, it breaks away completely from the visual and rids itself of any subjective factor that may enter into the assessment of the identity between two different elements. Symmetry in this sense is merely an *operation*, a movement whose point of ending is indistinguishable from its point of origin and this movement will be described—in accordance with the abstract methods of modern mathematics—as a permutation of numbers. In the case of bilateral right-left symmetry this operation is that of mirror reflection.

Reflection has the characteristic property of being *involutive*, that is to say that, applied once, it produces something different and applied a second time it reproduces identity. I shall have the opportunity of showing that even today there is such a thing as the "mystery of the mirror". This makes it all the more necessary that I should stress that the concept of the mirror image, as it is used by the mathematician when defining symmetry, has nothing mysterious about it but is precise and strictly mathematical.

On the one hand the operational definition of symmetry is prepared to account for the visible phenomenon of the mirror image, on the other it enables us to go far beyond the bounds of geometry "through the process of mathematical abstraction along a road that will finally lead us to a mathematical idea of great generality the Platonic idea, as it were, behind all the special appearances and applications of symmetry". Whoever takes the trouble to work his way through Hermann Weyl's book (a task which, though it involves no great mathematical knowledge, requires a good deal of mathematical sense) can travel the most important stretches of this road and, guided by the Ariadne's thread of symmetry, can penetrate the inmost arcana of theoretical natural science. The *group concept*, that most marvellous tool of modern thought, reveals itself here in all its universality and creative dynamism.

Thanks to the calculus which it provides, we can "count"

with symmetry operations just as well as we can with ordinary numbers. The reciprocal determination of the groups of operations and of their *invariants* permits us not only to recognize that our knowledge is "complete" but to fit the object of our knowledge into a new context of experience. Even in those abstract regions where anything like visualization has ceased to operate, the idea of symmetry thus still has a meaning and enables us to grasp something of the true structure of the world (cf. p. 169).

Nothing like this was destined for the second of the two concepts referred to, that of *polarity*. Polarity had played a certain role in so-called natural philosophy under the protection of the great name of Goethe. It proved, however to be too closely bound up with visual images, too much charged with feeling to endure the great cleansing process of mathematics which, if it succeeds in its aim, can lead to ideas of great unifying power, without forfeiting what in naïve experience was "really intended". Not that there is any lack of a mathematical concept of polarity, but the mathematical poles are nothing but particular symmetrical points and have nothing to do with the meaning which the naïve person, the philosopher and to a certain extent the scientist, attributes to this term. In the *Studium Generale*, though definitions for symmetry are suggested, there are none put forward for polarity. The botanist, Wilhelm Troll, seems to understand by polarity nothing more than the opposition (in space?) of different things, and we must ask ourselves whether this does not place polarity outside the more exact fields of natural science. Difference in principle—not to be confused with only approximate similarity—is not something that is amenable to logical-mathematical treatment. That is why Schopenhauer, who uses the concept of polarity, following Goethe, in his own *Farbenlehre* (theory of colour)—surely one of the strangest products of natural philosophy—seeks to confine it to a very special kind of difference. To be polar, two phenomena must substantially determine each other; they must further be opposites to each other *in specie* but identical *in genere*.[12]

Elsewhere[13] he speaks of a polar phenomenon as one which has "merely the appearance of an activity which falls into two halves which condition each other, seek each other and strive for reunion". He adds that "this falling apart mainly manifests itself spatially through a division into two directions of movement". In this way he hoped to explain the concept of polarity "without adding another faulty usage to the all too frequent ones that this concept has suffered in the period of Schelling's natural philosophy". One is inclined to doubt whether his effort succeeded.

Probably the concept of polarity can only claim to have a meaning in connection with certain particular theories of natural science, some of which are outdated. The concept was formed at the time of the agencies, fluids and other principles which neutralize each other, such as acids and alkalis, Perhaps it owed its origin to a misunderstanding of the physics of forces which Newton's system implies. "I had noted", says Goethe, "from my study of Kant's natural science that the powers of attraction and repulsion belong to the very nature of matter . . . this generates the basic polarity of all things which permeates and vivifies the endless variety of all phenomena." Or Schopenhauer, "Electricity propagates its self-division up to infinity",[14] and, referring to magnetism which is not produced by any galvanic apparatus but is fixed in steel and iron ore: "It is enclosed in a body, unchanging and till now unexplained: it is held fast within it as by a spell, like an enchanted prince."[15]

Charming as such analogies may be, modern science has little use for enchanted princes. Indeed it sets itself the task of removing the spell which binds them whenever it encounters them. It first succeeded in doing this in the case of electricity. The physicists were able to tear the electrical *dipole* apart. The magnetic dipole resisted them a little longer, i.e. the magnet retained till quite recent times its mysterious property of always having a positive and a negative pole, no matter how many tiny parts it might de divided into. Yet we are now engaged in freeing this last prince of physics from the spell by which he is bound: positive and negative magnetism proved to be intrinsically alike.

Perhaps biology now remains as the last sphere of knowledge in which polarity still means something.

Above all we cannot help being struck by the analogy between the characteristic property of the magnet that has just been described and the even more mysterious property possessed by living things, namely that of reproducing themselves by self-multiplication—that halving process in which each half in some strange manner becomes a whole. Is not the living form in this sense indivisible? Indeed "idealist morphology" sees in this a basic phenomenon, which one should and can accept quite calmly, in this matter following the advice of Goethe. "The blue of the sky reveals to us the basic law of chromatics. Here one should not search behind the phenomena. They themselves are the law." That the "blue of the sky" is only the beginning of the whole fascinating adventure of optics was something which Goethe and his contemporaries could of course hardly be expected to know.[16]

The present use of the word "polarity" has in fact a predomi-nantly biological colouring. The prototype of the "polar" is the cell, the germ with its animal and vegetable poles—two externally indistinguishable points which contain the opposing potentialities of the future individual. As Frey-Wyssling[17] points out, this "invisible physiological polarity in its dual ontogenetic and phylogenetic aspects still represents the basic problem of biology". Here, on the borderline between the living and the inanimate, at the most advanced post of research, modern science is engaged upon its most exciting battle, as it seeks for the "explanation" of biological polarity in the micro-scopic structure of living thinsg (cf. p. 106).

It is true that views vary as to the significance and probable issue of the battle according to the general way in which the morphological is assessed. Some people insist that Goebel's well-known remark, "The Morphological is what cannot *yet* be understood in terms of physiology", should be changed to "The morphological is what definitely cannot be understood in terms of physiology".[18] For them polarity is a basic phenomenon, the polar difference is "a potentiality present in the germ". Others —and these today are probably in the majority—do not scruple to get rid of the morphological categories wherever this is possible. For the scientist of this school the polarities he encounters are of more or less transient significance. Problems which still await their solution, they are, one might almost say, the princes whom the scientist is endeavouring to release from their spell.

Certainly both theoretical and experimental science have only been able to advance by taking symmetry and not polarity as their watchword. And this holds true also of the right-left problem. *"To the scientific mind"*, says Hermann Weyl, *"there is no inner difference, no polarity between right and left as there is for instance in the contrast between male and female or between the front and rear ends of an animal."* Elsewhere we read that "scientific thinking sides with Leibniz. Mythical thinking has always taken the contrary view, as is evinced by its use of right and left as symbols for such polar opposites as good and bad" (cf. p. 28).

Many will perhaps regard such questions as these as having been settled long ago. The supposed right-left problem, they will declare, had never existed anywhere save in certain idle pseudo-philosophical speculations; the scientist, at any rate, or the man used to scientific ways of thought, has, nothing to do with right and left; the right and left sides of an equation are interchangeable, they are simply "equal" like the right and left sides of a scale, and Aristotle had already observed in his Physics that when I turn around, what had previously lain on my left now lies on my right. All this is no doubt true. But we only need to think of Kant to be reminded that nothing can be accepted as self-evident. For less than two hundred years ago, at a time when people already knew very well that the men of the antipodes did not walk on their heads and that there was therefore little reason for speaking of directions in space, he wrote his treatise *On the First Ground of Distinction between the*

Regions in Space, which might well have been entitled "On the First Ground of Distinction between Left and Right in Space". In the pages that follow we shall encounter a number of well-known scientists and philosophers for whom there was a right-left problem, which means that like Kant they could not completely free themselves from the right-left myth.

Before we go on to that however, I propose first to examine the second part of Weyl's statement that mythical thinking brings right and left into a polar opposition of the same kind as good and bad. I will therefore pose the question: is right better than left?

II

IS RIGHT BETTER THAN LEFT?

1. *Manichaeism in language and custom*

Until a comparatively short while ago "mythical thinking" was regareded as something entirely different from our own, something that obeyed its own special laws, such as those of analogy and participation, something that formed a closed homogeneous whole and was classified by us as "primitive". In these circumstances it is not surprising that almost every publication which touches upon the *evaluation* of right and left— for the most part medical and psychological treatises on the left-handed—advances the view that such evaluation is as consistent in character as mythical thought itself. When dealing with the two terms "right-left", regarded as a pair of polar opposites, right was unfailingly identified with good and left with bad. This view is supported by such an impressive number of examples that even the most enlightened mind begins to wonder if right is not really better than left.

In recent times, however, considerable changes have taken place in our ideas about mythical thinking. Indeed no less a person than the actual creator of the concept, the sociologist L. Lévy Bruhl,[1] was one of the first to confess and correct his errors in his posthumously published *Carnets*—a remarkable example of scientific self-criticism. For the student of today there is no such simple contrast between the granite monolith of primitive pre-logical thinking on the one hand and our so quite different occidental rational thinking on the other. At the same time the former now seems much less strange to us.

This change of attitude has caused us—to a degree hitherto quite unknown—to take seriously such mythical thinking as could be found in foreign and exotic cultures and also in our own folk-lore. The scientific enquirer today takes account of the persistence of certain mythical images and is careful not to exclude the possibility that there may be some reality corresponding to such primeval intuitions, a reality that he may be able to grasp. On the other hand he will mistrust those who tell him, as they continually do, that for mythical thinking right is *invariably* better than left. My introductory chapter gave some idea of the incredibly fine variations in the concept of pairhood in mythical thinking. It is obvious that ideas which we have called "non-Manichaean" cannot be reconciled with a preference for the right side over the left. And after all why should the left not be "better" for some people than the right? If we focus our attention on the alleged "exceptions" and deviations from the norm, we are surprised to find them so numerous and significant that the "dogma" of right being better than left emerges, to say the least, very badly shaken.

There are many examples in which we can see the emergence of the Manichaean good-evil pattern in language. Thus in German *rechts* (meaning on the right side—old German *rehts*, from the Latin *rectus*) is connected with *Recht* meaning law; the adjective *recht* means conforming to the rules, right as the opposite of wrong (as in the right way) or right in the sense of straight, as in right angle. On the other hand, *links* meaning left (old German *lenka*), has a pronounced pejorative sense as in *linkisch*, meaning clumsy or *linke Seite* when used in the sense of "back".

It is much the same in French. *Droit* (from the Latin *directus*) in the sense of right as distinct from left, is synonymous with *le droit* in the sense of law, with *droit* straight or right in the sense of correct (*le droit chemin*). We find it again in the expression adroit which English has borrowed from French. Whereas *envers*, meaning the back or reverse side has nothing to do with left, the front side is nevertheless referred to as *endroit*. *Gauche*, left (from the old French *guenchir* = to make a detour, to bend) has the sense of clumsy. We must further note that one of the two Latin words for left has produced no word for left in French —if we disregard the rare and purely scientific use of *senestre*, the opposite of *dextre*, and *senestralité*. As against this however we have the French *sinistre* with the same meaning as the English sinister, and also the identical noun meaning a catastrophe.

In Italisn *destro*, meaning the right side, has the meaning skilful, though the three words for clumsy have nothing to do with *sinistro*, the Italian for left. Here too however *sinistro* has the meaning of disaster. It is also worth noting that *mancino* not only means a left-handed person. It also means a thief!

The examples, which are always cited as illustrating the pejorative use of left and the laudative use of right, will naturally only convince those who are inclined to accord decisive importance to the linguistic argument. A very ancient problem reappears here, one which Plato had already posed. Do things really have their "right" names, as Kratylos naïvely believed? And could one infer a natural kinship from a linguistic one? "Kratylos here, O Socrates, declares that everything has a name that is suited to it by nature." As we know, Kratylos had to submit to the arguments of Socrates. Only the poet believes that a star could have no other name than "star" and for him alone "seagulls all look as though they were called Emma". The scientifically trained person however never doubts the essentially conventional character of language. He knows that a child, left by itself and in the absence of that "other" who comes to an understanding with it as to how these things are to be called, would never learn to speak at all and that the story told by Herodotus about King Psammeticus is only a pretty fable. According to that story, it will be remembered, two children whom the king had caused to be set apart in the wilderness in order to see whether they would develop speech and if so what language they would use, called out *bexos* when they saw him, which means bread. But, as I have said, the story is pure fancy and, till the contrary is proved we must assume that right, left, good and bad and everything else in the world have the names which were given them by ourselves. That the reasons which cause us to give a thing a particular name are often much more banal than might be supposed, can be seen if we study a language with an ideographic script, a language that is to say in which there practically is no etymology. In Chinese the directly suggestive ideogram for right (Hand + mouth) shows us as soon as we look at it that for the right-handed Chinese the right hand is the hand that carries food to his mouth. It thus remains for the left to be the side of magic and the arts, and so left is designated by the combination of signs Hand + Carpenter's rule. Which of the two sides is the better or more important we are left to decide for ourselves.

These remarks are not intended to invalidate the examples here adduced to prove that right in our speech is accounted better than left. They do however remind us of the fact which we sometimes tend to overlook that there is no such thing as a "natural" language; rather are there many different languages and language groups which would all have to be studied in any linguistic examination of right and left.

Besides it is precisely the example of right and left that moves us to call in question the value of even a very thorough examination, if it is only concerned with the immediate present and neglects the changes in meaning that have taken place in the course of history. To take but a single instance, one of the Latin words for left, *laevus*, only acquired the meaning "clumsy" in Imperial times; the pejorative use of the word had not therefore diminished in the course of time as might have been expected, but had on the contrary been intensified. But the historical method of approach discloses something even more remarkable. The meaning of this same word *laevus* kept swinging between the two extremes of favourable and unfavourable according to whether the Romans were following the Etruscan or the Greek rites of augury. For when they followed the Etruscan rite the Romans turned their faces to the south and so had the west on their right; when they followed the Greek rite however, they turned north and so had the east on their right. So as we go back in the past we find that there are variations in the association of ideas. Moreover, the precariousness and insufficiency of any purely linguistic approach becomes all too evident. Compared with the ritual, which in this matter is the really determining factor, language seems a mere epiphenomenon, or, to put the matter a little differently, behind speech there is always "something else".

In the example to which I have just referred, this "something else" was the point of the compass. One might indeed ask oneself whether the *primary* association was, not right=good and left=bad, but east=good, west=bad, left and right being secondarily equated with bad or good according as those who "created" the language had the east on their right or their left.

It would then not be the sides but the points of the compass that had a "natural" significance, or a "natural" value. Of course I am here assuming that the side of the rising sun is the "good" side and that this conviction holds good not only *sub specie* of the freezing men of the north but among all mankind: let us not forget that the Hell of the Tibetans is a *cold* Hell.

In many languages we can clearly discover a cosmic-geographical origin of the words for left and right. To the Chinese east=left; the Egyptian uses the word "face" for south and for north a word that is related to the expression used to designate the back of the head. The word used to describe "east" and "left" is the same and correspondingly the same word is used to designate "west" and "right". In Hebrew *yamin* (right) and *sem'ol* (left) are used for south and north, sometimes we come across quite different derivations for the same words, derivations that have nothing to do with the points of the compass. For those who defend the "weapon hypothesis" for the explanation of left-handedness (cf. p. 119) left denotes the side of the body that has to be protected, the "heart side" or the shield which is designed for its protection; according to their interpretation the Hebrew *sem'ol* is "that which is concealed or covered", and the Cymric word for left, *asw* is said to come from *aswy*—shield. When such etymological reasoning breaks down, efforts are made to associate the left hand, which is said to stretch the bow while the right hand holds it, with all manner of expressions denoting crooked, bent or bending: reference has already been made to the old French word *guenchir* which has this meaning. There is also the Latin *laevus* which is supposed to come from a hypothetical root *lei*=to bend. Further there is a particularly daring hypothesis which derives *sinistre* and *sinister* from *sinus*=bow. Oddly enough there are no such etymological fancies associated with the word right.

The above remarks will perhaps show us that it simply will not do to look on speech as though it was something independent of human doing and being. Even so, custom, habit, popular beliefs and religion tell us nothing to contradict what we have learned from human speech. Innumerable examples testify that the right is the privileged side. Thus the right side of the body and especially the right hand hold a privileged place. It is the right hand that is extended in greeting and it is the right hand that is used in eating, it is towards the right that the cup moves in Plato's *Symposium*. The right hand is raised when swearing an oath, and a left-handed marriage is no proper marriage at all. All ritual acts—and this is as true of Jews and Mohammedans as of anybody else—must be performed with the right hand, whether it be the laying on of a hand in blessing or the cleansing of a leper.

The use of the left hand is accounted discourteous. The left hand is charged with the unpleasant duty of administering the *coup de grâce* to a wounded man (which is why in the Middle Ages even the dagger was designated as the "left hand"). There are even people who speak of it as the "hand of the privy". Similarly getting out of bed with the left foot is accounted a bad omen or is regarded as the cause of ill humour. "Concerning Divine Augustus," says Pliny, "we know that he had put on his left shoe before his right one on the day on which he almost became the victim of a rebellion."

God himself appears to share in this discrimination between left and right. Does it not say in Matthew v. 25 . . . "And He will set the sheep upon His right hand and the goats upon His left. Then shall the King say to those upon His right, 'Come, ye blessed of my Father, and inherit the kingdom prepared for you from the beginning of the world' . . . Then shall He also say to those on the left, 'Depart from me, ye accursed, into everlasting fire prepared for the Devil and His angels.' " Of the two thieves who were crucified one on each side of Christ, it is the one on His right that gets into Paradise after all.

The symbolism of the body corresponds to the symbolism of space whether it be great or small. Students of religion such as Mircea Eliade show us that *homo religiosus* feels the need to live in a "sacred space" and that hierophany reveals to him a firm centre in the limitless *apeiron*, a navel or heart of the world whence all the quarters of the heavens extend. "In order to live in a world one must give it foundations".[2] And so in all his "foundations" he repeats the cosmogeny and creates on earth the miniature representations of cosmic space: in the temple or sanctuary, in whole cities, be they Jerusalem or Roma Quadrata, in entire countries such as Palestine or Egypt, and even in the smallest village. In all these microcosms right and left have a special significance. Indeed in occidental cultures a privileged place is given not only to the right side of the body but to the right side of any room or space. People enter a temple or a church by the right door. Right is, generally speaking, the side of the Heavenly powers, left that of the diabolical and demonic, above all of the Devil himself, who naturally enough uses his left hand to fiddle with. Where it is the custom for a newly married pair to walk around the altar, they do so moving towards the right. There is also the example of the Basque theatre, with its very ancient tradition, which shows an interesting transition from a sacred to a profane building. Not only the Devil but all wicked persons, including the Communists [*sic*] make their entrance through the left-hand stage door. Whether the different significance of the two sides of the stage in a modern theatre—in French they are *coté cour* and *coté jardin*—owes its origin to mythical-religious considerations or to aesthetic ones such as the "law of sight direction" is difficult to determine. Unlike the corresponding symbolism of the body, the right-left symbolism of space contains an arbitrary factor in so far as right and left in space are not given but have to be newly "defined". Thus the statement, "Birds coming from the left promise good fortune", and "Birds coming from the right promise good fortune", have a special meaning for each person who utters them (the Greeks who opt for the second alternative believe that right is better than left); but to which side of space the right or left side of the body corresponds will depend on the position of the observer and the direction in which he is looking. The Greek augur who looks to the north because Zeus lives on Mount Ida in the north of Greece, has the west on his left. Numa, so Livy tells us, took up his position in the north and looked southwards because as emperor he wanted to take up the position of the god whom he represented, and so he had the east on his left. In much the same way the Egyptian looks southward when he wants to take his bearings, for to the south lie the sources of the Nile which are also the source of his prosperity.

A second "arbitrary" factor was introduced through the circumstance that the points of the compass, as I have already pointed out, have no "natural" values. Thus for the Old Testament the east is the country of the sun-worshippers, the enemies of Israel, and it could therefore never promise good fortune. In view of the many possible permutations and combinations of right, left, good, bad, east and west, it seems hopeless to attempt an interpretation of such an apparently innocent phrase as, "Birds that come from the left promise good fortune", and one can understand why, Cuillandre devoted a great part of his five-hundred page thesis on *The Significance of*

Right and Left in the Homeric Poems to this problem of augury interpretation. This author seeks to prove, in the spirit of "Aryan mysticism", that the Greek consulting an oracle and the Breton fisherman speaking of *ar mor dehou*, the right-sea, and *ar mor klei*, the left-sea, are both basing themselves on the same principle of orientation. He will not allow that Latins, Greeks and Celts may have different rites; there can only be different aspects of the same ritual, in which the right side, which in his view *is* the good side, must necessarily appear as such. He therefore introduces a new variable: the *interpretations* of those who consult the oracle. Let us by way of illustration assume that a Latin and a Greek both turn their faces towards the east, towards the rising sun, and an eagle coming from the north flies over the *templum* and passing over the east flies south. The Latin according to Cuillandre, would say that it comes from the left, *laevus*, and brings good fortune because it flies in the same sense as the movement of the sun, i.e. *to* the right. The Greek would say that its position was towards the right, *dexios*, and promised good fortune on that account. So each in his own way could still regard the right as the better side. What this example really illustrates is the frequent preference entertained by scholars for the right side. It would have been simpler to assume that some peoples do prefer the left and to accept the word of the elder Pliny when he says of these, *Laeva prospera existimantur*. And that is just what I propose to do.

2. *Right and left in relation to the categories of sex*

J. J. Bachofen the Swiss legal and cultural historian, in his famous work *Das Mutterrecht* (Matriarchy),[3] published in 1861, adduced innumerable examples of "reversed" values being given to right and left. He connects this—a very original idea—with the social and cultural phenomenon of *matriarchy* which he himself had discovered. The reversed social system leads logically to a preference for the "other" side. The assumption behind this is of course that the equation female=left holds good—a point which for the moment I will not argue.

Just as we ourselves honour the right, Bachofen suggests, so those peoples for whom the woman and mother are especially holy honour the left side; it is honoured as "the symbol of the substance of motherhood, the expression of the female substance which gives birth, nourishes and multiplies in all the diverse expressions of its activity". The highest development of this cult was attained in the Isis cult which gave precedence to the Isis principle over that of Osiris, to night over day, to the mother over the son. The Isis procession is headed by the priest who carries an image of a left hand, which, as Bachofen the jurist points out, was also called *Justitiae manus* and which symbolized a mind impervious to *calliditas* and *sollertia*. So for the gynaeco-cratic conception of the world *justice* is associated with the left side and is symbolized by the figure two, which Pythagoras himself had regarded as "female". Again agriculture (Mother Earth) naturally finds its symbolic expression in the left side. Libyan husbandmen cut the hair on the left side of their heads "in honour of chthonic motherhood".

These two basic themes, left=justice, left=Mother Earth, are developed in *Das Mutterrecht* with a wealth of examples, but it would go beyond the scope of this book to list even a fraction of them here. Some of them seem to me rather forced. Who, for instance, would regard it as a victory for the ideas of matriarchy that Jason lost his left sandal in the marsh and owed his victory to this mishap? (It was with the left foot, be it noted that Mopsius trod on the deadly snake.) For all that Bachofen deserves credit for having freed us from our occidental prejudice and proved to us that left is occasionally—even frequently—"better" than right.

Now that ethnological research is freeing itself more and more from "the compulsive idea of religious dualism" it is constantly finding instances of special value being attached to the left side. In his book about the Forest Indians of North America, for instance, to which reference has already been made, Werner Müller, says this of the Delawares: "The oval path on which the dancers move around the two fires is swept by the Ashkas with turkeys' wings which they hold in the left hand, *for the left is holy, the right unholy.*" He adds a remark which reminds us both of the connection of these ideas with the sex categories and of the ambiguity which can characterize the process of interpretation in such matters. "Male and female servants begin to sweep from opposite ends. The male servants starting from the west and moving towards the east while the female ones start from the east and move toward the west." Numerous instances are also cited by Garnet in a chapter of *La Civilization chinoise* which is entirely devoted to this question of right and left, of honour being accorded to the left side or the left hand in ancient China (cf. note 8 to Chapter I).

Although Bachofen set research on the right track, his equation of left with female has in the final analysis proved incorrect. The Chinese example, which Bachofen neglected for the study of Plutarch and Apuleius, shows this: for the Chinese the left side was Yang, that is to say *male.* So the millions of people who in China, unlike ourselves, honour the left side, are thereby honouring the male principle, not the female. We can no more associate this sex or that with right or left than we can associate some specific point of the compass. In much the same way matriarchy does not necessarily bring with it a higher esteem for the left nor does patriarchy produce a higher esteem for the right. There is no *"left-culture"* that could be considered a mirror-image of a *"right-culture"*. One has only to look at the social and legal relationships to dismiss any such idea. For instance, there is nothing in the patriarchal society corresponding to the uncle, the mother's brother in the matriarchal society. Of those students (the reasonably serious ones) who have based their systems on the concepts of right-culture and left-culture, one can only say that they have fallen victims to the "illusion of the mirror", an illusion to which we shall be making frequent reference (cf. concluding chapter).

We can be pretty sure that in principle all imaginable permutations and combinations are possible and that we should probably be able to find examples of them. There are peoples who equate left, female and north, and consider such a combination to be bad; there are peoples like the Egyptians for whom left is identical with female, east and good. The choice of symbols, is arbitrary. Not, or course, in the sense that the individual is free to choose them—the society in which he lives chooses on his behalf—but that society has a certain freedom in making this choice, which is not imposed by nature. In particular the symbolic use of right and left is something which has evolved and taken shape and, like most things of this kind, is the joint product of necessity and chance, a mixture of chaos and order, and is comparable to those sequences of random elements in mathematics which are known as Markoff chains.

Little remains today of Bachofen's great theoretical edifice save a mass of material facts—and even these are in part disputed—facts from among which others can pick out whatever they happen to require. Yet, like Kratylos, we humans have always been eager to hear the original language of things and that is why it was in fact the *theories* of this jurist from Bâle which brought him a large body of followers because those theories happened to satisfy that particular need. An idea of Bachofen's which seems to have had a particularly wide appeal was that the older cultures were matriarchal and that the female-chthonic left-cultures only gave way to their opposites, the male-uranian ones at a later stage, which was why—to take one example—the Delian *Keraton* was originally constructed of left horns and only later of right ones. Richard Kobler[4] has advanced the view that there is not only a symbolic but a "real" connection between female and left. According to this author, who is as stimulating as he is unreliable, the gynaecocratic cultures were left-cultures in a very practical sense, since they were the cultures of left-handed people. This curious theory is wholly dependent on the assumption that in prehistoric times there were proportionately more left-handed people than there are today. Kobler endeavours to prove this in his book *Der Weg des Menschen vom Linkshänder zum Rechtshänder* (The way of Man from Left-handedness to Right-handedness). He based his theory on the researches of C. de Mortillet on hand axes and other prehistoric stone implements, having particular regard to the cavities made in these implements and to the "feel of the grip" when they are grasped alternately with the right hand and with the left. He claims to be able to determine, as one can do with, say, a pair of scissors, whether they are designed to be grasped with the left hand or the right. According to Kobler and de Mortillet the statistics show a preponderance of "left-handed" implements. So far as I am aware, however, these investigations have not produced any convincing results.

The same is true of the perhaps even more interesting studies of cave drawings. A greater frequency of left-handedness has been deduced from two sets of facts: the frequency with which such figures as those depicted in the hunting scenes are shown using their left hand or their right, and the tendency to show the profiles of men or animals from the left or from the right. It is a well-known fact that right-handed people prefer to draw left profiles; indeed R. Zazzo states that 70 per cent of all profiles drawn by children face left. The left-handed Leonardo on the other hand drew right profiles almost exclusively.

Kobler is quite unable to explain the reason for this change-over from left to right which took place in those distant days, nor can he say how it was effected. Did some mutation produce a sudden and radical change? Did some authoritative father-like super-ego which commanded such a change evolve out of matriarchal society for reasons which we cannot fathom? One thing is certain: Kobler's theory is contradicted by the simple fact that the Chinese, though their culture is probably derived from an ancient matriarchal one, are as right-handed as we are.

It is remarkable that the positive value placed on the left is never as absolute as that placed on the right. Those who give preference to left over right do not simply turn the whole pattern upside down. They do not choose left instead of right but always to some extent left *and* right. One gains the impression, therefore, that the opposite of "right cultures"—we must of course exercise a certain caution in using that expression—is not "left cultures" but *intermediate or balanced cultures*, which are less concerned with the triumph of a principle than with the harmonious balance between a number of principles which are complementary to each other. Bachofen had already praised the "mild humanity" of these cultures which he regarded—let us never forget this—as typically female. Presumably a more tolerant and conciliatory philosophy, which was in fact the philosophy of a minority—and the left-handed are in fact such a minority (cf. p. 134)—may have found expression in them. Even night, even death find their place here instead of being eliminated—in the Manichaean manner—by the eschatological process. In the cyclic conceptions of the *Indians* the dancing Shiva holds in his four hands the symbols not only of life but also of death, the necessary condition of life.

It is in this sense that we should understand the findings of the Dutch ethnologist Albert C. Kruyt concerning the associations life-day-right and death-night-left among the inhabitants of the middle Celebes. The dead, according to the beliefs of these tribes, do everything "in the opposite way to the living. What the living do with their right hand the dead do with their left—or rather, what is right for the living is left for the dead." "Because he is dead, he must choose the left path which is for him the lucky one, on his way to the kingdom of the dead." The same principles govern the burial rites. Food and utensils for the dead are arranged in such a fashion that he can easily grasp them with his *left* hand. Rice for the dead is only held in the left hand. It is thus only logical that the left is under no circumstances allowed to touch rice intended for sowing. Ernst Siegrist remarks in his *Zur Händigkeit des Menschen*[5] (On the handedness of Man) that this idea does not imply a value judgment such as right = good, left = bad. We are here concerned simply with a pattern of things,which puts everyone in the world whether living or dead, into his right place, and which lays down what is right conduct for him. The "overwhelming force of such tribal mythology" convinces Siegrist that "it is primarily the general outlook (*Weltanschauung*) of the person concerned which determines whether an action should be performed with the left hand or the right, and no regard is had to the complexity of the process which the task involves or to the personal predilection of the person engaged on it". Here is a weighty argument in refutation of Kobler's theory.

3. *The Chinese problem*

In Werner Speiser's book *China—Geist und Gesellschaft* (China—Spirit and Society)[6] there is an illustration of a bronze vessel, with a lid, dating from the second century. Its two "display sides" are formed by the two enantiomorphic profiles of an owl which has thus been divided into two, the two parts being fastened together back to back. Speiser remarks that this bird which is often found on sacred Chinese bronzes, because of all birds it is the embodiment of darkness, is chosen to symbolize the air, and is contrasted with the symbol of the earth—which incidentally always appears alongside of it. The effect is that the polarity earth-air/light-darkness appears to be broken through, for the fact is that Chinese logic avoids antitheses and loves correlatives. He expresses the fear that his interpretation may be regarded as over-ingenious. Actually, however, it harmonizes very well with the conception of Chinese philosophy as a philosophy of the centre, of harmonious compromise. To this I might add that on the owl vessel it is not only the polarity earth-air/light-darkness that is broken through, but, thanks to the spatial arrangement itself—this owl unquestionably reminds us of Plato's myth and of the beasts of the Apocalypse—the *polarity of left and right*.

The Chinese problem has caused some confusion where—as

happened with Bachofen—it has not simply been ignored. Kobler declares that for the Chinese, as the representatives of a very ancient gynaecocratic "left-culture", the left side has always been regarded from time immemorial as the better one, this being in accord with a famous saying of Lao-Tse which he translates as follows:

Good fortune dwells on the left/Ill fortune dwells on the right/ The soldiers stand on the left/The leader stands on the right/ News of war—news of grief/Death of human beings—source of tears/ Victory through force is suffering.

To this Siegrist rightly remarks, "It seems to me to be somewhat forced, one might even say that it betokens a lack of understanding, when people endeavour to reduce the whole content of this saying to the relation: right=bad=right-handedness, left=good=left-handedness."

Before endeavouring to delve deeper however, let us replace this quotation which, even by Chinese standards, seems somewhat brief, by Richard Wilhelm's translation, into which I have only worked a few words of Waley's rendering, which seems to me rather clearer.

"In hours of good fortune the left is the place of honour/The right is that place in hours of misadventure/In war the lieutenant stands on the left/His commander however on the right/When a great number of people die/ It is fitting to wear mourning/ And one does well to liken a victory to a burial."

The first thing that strikes us about the above is that the author of these verses obviously prefers being a lieutenant (Kobler even says a soldier) to being a commander, a mental attitude which is comparable to that behind the words quoted by Needham from *Tao Te Ching*: "Therefore the sage holds the left tally (the less honourable or inferior side, i.e. takes the part of the people), and does not demand from the people the impossible." Needham makes the following comment: "My friend, Dr. O. von der Sprenkel has pointed out to me evidence from *Chhien Han Shu* that in Chhin and in former times the poorer people lived on the 'left' side of the village—on the

wrong side of the railway tracks, as some might say."[7] In both
cases the Chinese sage obviously honours in the left the side of
the weaker, of those who are more in need of help, whose cause
he deliberately makes his own because for him it is the "better"
one. One is reminded of Sartre's ethical conjunction to take the
side of the weaker and also of the words of Jesus "It is more
blessed to give than to receive", and it is hardly surprising to
hear that the ancient Chinese ritual for gifts actually lays it down
that giving should be done with the left hand and receiving with
the right.

Moreover, this saying of Lao Tse seems to show that the
significance of left and right, could change according to the
prevailing circumstances and protocol. The left is not always the
side of honour. Granet cites numerous examples which confirm
this impression. *Etiquette*, which in ancient China had to take
account of the infinitely complicated interactions of different
categories, sometimes advanced the left to honour and sometimes
the right. In eating the right hand is used; the second finger on
the right hand is not the index finger—to point is dangerous and
is forbidden—but the eating finger. In greeting it is considered
proper for boys to hide the right hand under the left, for it is the
hand that kills. Girls hide the left hand under the right. But in
times of mourning when Yin and the right gain the ascendancy,
men give their greeting after the manner of women. The left
shoulder is bared when attending a joyous ceremony and the
right when punishment is expected. When men swear brotherhood they grasp each other's right hand. In the case of blood
brotherhood, the blood that is exchanged is taken from the right
arm. But if, as happened occasionally in earlier times, an oath
was to be strengthened by the sacrifice of a prisoner, blood was
taken from the victim's *left* ear, which on such occasions counts
for more than the right. One also held the prisoner's fetter in the
left hand (cf. p. 112). The list could be continued.

Granet explains these apparently contradictory rules as effects
of the social and political origin of the categories left and right.
"It has long been observed", he says, "that the whole of Chinese

wisdom pursues political objects." . . . "The idea that seems to permeate the whole edifice of learning is that human thought does not aim at pure knowledge but at civilizing action and that its task is to secrete an effective and total order. There is therefore no concept that is not linked to some specific attitude, no doctrine that is not associated with some recipe for the conduct of life. There is no distinction between the order of the universe and the order of civilization.

Granet attributes the changing significance of left and right to their combination with the sex categories and with the categories of above and below. This is clear, for instance, in the case of the vassal who at court honours the west and the right, but reverses all the rules of etiquette when he is at home and is once more the master. He also points out the double dissymmetry which the distinction of above and below introduces both into the microcosm and the macrocosm. In the upper part of the body, in the eyes and in the ears, Yang, which is correlated to Heaven and the left, is supreme, while in the lower part, which is nearer to the earth, it is Yin and the right. But the world as a whole is to some extent out of true and contains, as Weyl would say, an inward twist. Granet tells the myth of Kong-Kong which explains how, as a result of some mighty cosmic event, a lack of Yin and of Yang began to be felt in the corresponding parts both of the body and of the world as a whole. Nü Kua, whom we have already encountered, was unable completely to repair the world that the wind spirit Kong-Kong had devastated. Since then the square earth carries the sky and the round sky holds the earth in its grasp, but both have shifted somewhat towards one another!

The most pleasing picture of a harmonious balance and interaction between left and right is furnished by the mythical accounts of "half" heroes. The most outstanding heroes of ancient China, Yu the Great and T'ang the Victorious, are partly right-handed and partly left-handed. They are, as it were, geniuses of the right or geniuses of the left, being masters either of rain or of drought, entirely devoted either to Yin or to Yang.

Hence people are prone to declare that they are unilaterally paralysed, when they do not actually reduce them to the left or right halves of their bodies! The founders of two succeeding dynasties—these heroes are of course founders of order and civilization—must be inspired either by the genius of earth or by the genius of Heaven, but neither the hero himself nor the dynasty founded by him is better or worse because the hero is left-handed or right-handed and so animated either by the virtue of Heaven or that of earth. These virtues are complementary. They must alternate with each other in their work.

Moreover, they permeate and impregnate the perfect *saint* during the course of his life, first the one and then the other. First he is a minister with active duties. He develops his gifts as he deals with the details of earthly concerns. When afterwards he becomes a ruler he is solicitous about Heaven and only lives to develop in himself that higher efficacy (Tao) which is superior to all efficacy in matters of detail.

4. *Harmony and balance in the West*

The question arises whether such harmony and dynamic symmetry can derive entirely from the "wakeful bright stillness of Tao". Certainly this attitude fits in with the whole structure of the ancient Chinese world, a world of "self-sufficient order", without God or gods. For gods are frequently intolerant; the scales—for the Cabalist the symbol of a movable equilibrium— become in their hands the symbol of a rigid dualism. On the other hand, there is little in the history of western thought—or in western history as such that escapes the "good-evil schizophrenia", which, as Needham shows in his profound analysis of the law concept, leads logically to such extravagances as the trial of animals, the conviction of a pig for murder or of a cock for the "heinous and unnatural crime" of laying eggs. If,

however, we take left-right symbolism as an "index" of the
extent of Manichaean derangement, we come to a more
optimistic conclusion; for even in European thought there are
many examples of a different kind which can be set against the
examples of discrimination against the left quoted above.

Let us start by opening the *Bible*. It is, I suppose, of relatively
little importance that in the *Old Testament* left-handedness,
which is often an advantage in modern sports such as tennis and
boxing, is considered "practical" in armed combat. We are told
in Judges 20 that among the 26,000 children of Benjamin who
drew the sword were "700 chosen men left-handed". Ehud, the-
God-sent saviour of Israel, was also left-handed (ibid. 3.15) and
used this advantage to kill Eglon, the King of Moab in a highly
questionable manner.

On the other hand, it is worth noting that in the Old Testament text right and left themselves occur very rarely. Apart
from the passages already referred to, mention of right and left
is made principally in reference to blessing and to the stereotyped
repetition of the cleansing rules (Leviticus 14.14 on the cleansing
of leprosy in man and house). In both cases there is a direct
connection with right-handedness.

There is no mention whatever of right and left—and this
omission is strange—in the directions for the building of the
tabernacle of the ark of the covenant, or of the altar for the burnt
offering, though the basic significance of the holy space and its
microcosmic representation (cf. p. 34) would lead us to expect
that if a distinction between left and right occurred anywhere,
it would be in this context. Yet in all these minute directions
concerning knots and hooks there is nothing dealing with a
direction except what is enjoined concerning the placing of the
boards. "Twenty shall stand against the noonday and twenty
against midnight." The same is true of Aaron's ephod, the design
of which is very exactly described. As to Aaron's breastplate,
this was to be a span in length and a span in breadth and it was
to be filled with *four* rows of stones.

In the number *four* which occurs so frequently, particularly in the prophetic texts, we can quite definitely recognize a sign of the suspension of polarity (cf. p. 17). The most impressive text of this kind is the vision of the prophet Ezekiel, which, incidentally, Jakob Boehme interprets in just this sense.

"Now as I looked at the living creatures, I saw a wheel upon the earth beside the living creatures, one for each of the four of them. As for the appearance of the wheels and their construction; their appearance was like the gleaming of a chrysolite; and the four had the same likeness, their construction being as it were a wheel within a wheel. When they went, they went in any of their four directions without turning as they went. The four wheels had rims . . . and their rims were full of eyes round about" (Ezek.1.15).

For the benefit of anyone who still doubts that, as in the book of Genesis and in the Symposium, we are concerned with a myth about twins, a myth referring to rotation symmetry, let me quote the description, given earlier in the chapter, of the four creatures that accompany the wheels: "Each had four faces, and each of them had four wings. Their legs were straight (!), and the soles of their feet were like the sole of a calf's foot; and they sparkled like burnished bronze. Under their wings on their four sides they had human hands. And the four had their faces and their wings thus: their wings touched one another; they went every one straight forward without turning as they went. As for the likeness of their faces, each had the face of a man in front; the four had the face of a lion on the right side, the four had the face of an ox on the left side, and the four had the face of an eagle at the back. . . . And their wings were spread out above; each creature had two wings, each of which touched the wing of another while two covered their bodies" (Ezek. 1.6).

Later Jewish religious writing strengthens my impression that the Old Testament did not discriminate against the left. Along with a number of examples showing preference for the right, I take from the very detailed paragraph in the *Jewish Encyclopedia* on the evaluation of right and left a number of practical directives and interpretations of the Old Testament that support my own view. In the wedding ceremony the bridegroom stands on the left of the bride under the canopy though that is something which can be interpreted in various ways. In another ceremony the palm branch Lulam is carried in the right hand while the no less important pomegranate, Etrog, is carried in the left. The Zohar describes the Torah, the written law, as "right", and the verbal tradition, which also plays an important role, as "left". In the Cabbala there is a noteworthy instance of the left being honoured on ethical grounds. Phylacteries should be worn on the left because this is the side nearest the heart and more directly in the service of God. "Is there a right and a left up there?" asks the student in *Midrash*, referring to a test in the first Book of Kings. "I saw the Lord sitting on his throne and the Heavenly hosts were with Him upon His right hand and upon His left." The teacher answers: "But there were defenders on the right and accusers on the left." Clearly the student is here adopting the attitude of Luther: *Dextera Dei ubique est* (cf. p. 51).

What seems to me particularly illuminating is that in the Zohar our world is called the right one and the world to come the left. The *Jewish Encyclopedia* comments in this connection: "It is curious that in the Cabbala the left side represents a higher and more developed state." There seems indeed to be a tendency here to endow left and right with a significance which is the opposite of that which is customary in modern sciences of expression. Whether by accident or because of some more profound reason, this tendency is quite in line with our modern political symbolism.

In the New Testament mention of right and left is even less frequent than in the Old and, apart from the text in Matthew to which I have already referred, important though it certainly is, it would be difficult to make out a convincing case showing the existence of any kind of preference for the right. Does not the mother of the sons of Zebedee say in this very book, "Let my two sons sit, one upon thy right hand and one upon thy left", to which Christ replies, "To sit upon my right and upon my left is not for me to give but is for those for whom my Father hath prepared it."

As in the Old Testament so here too, the later development of right-left symbolism is interesting. There is often a distinct Manichaean tendency in Christian religious art. For instance, hard and fast rules grew up for the representation of the Crucifixion, distributing good and bad and apportioning honour and dishonour to different sides of the Cross. Often, indeed, the introduction of allegorical or historical personalities whose position is not fixed by the biblical text—as is that of the two thieves—leads to disputes concerning their rank. John must ultimately take his place on the left of the Lord so as to make room for the mother of God on the right and so on. A particularly perfect example of this right-left symbolsim is to be found in a miniature of the Hortus Deliciarum. Mary, the Centurion, the penitent thief and the Church, riding on a four-headed apoca-lyptic beast are on the right; to the left of the Cross are John, the bearer of the sponge, the impenitent thief and the synagogue riding upon an ass.

In the figure of Christ Himself however, the left ranks presum-ably as the equal of the right. Thus in Chartres Cathedral Christ holds the Gospel in His left hand and blesses with His right. Similarly, the Gospel is read on the left side of the altar. All representations of Christ triumphant are distinguished by their perfect symmetry which overcomes the right-left polarity.

According to the *Dictionnaire d'Archéologie Chrétienne* by Cabrol and Leclercq the right only became the side of honour for the Church under Nicholas IV in 1288. In the early days Christians had little time for symbolic lucubrations and preferred to occupy themselves with describing and moralizing; it was only in later years that this changed. I believe however that the choice of symbols often carries the implication of profound ethical and ontological decisions. The works of many great Christian thinkers bears witness to this. Jakob Boehme, for instance, despite—or perhaps because of the fact—that "all things are double", thinks it important that we should recognize that there can be no directions in the divine: "Stands Christ or sits He (at the right hand of God)? . . . Behold, he sits within Himself and

stands within Himself, He needs neither seat nor chair, His
power is His chair; there is no above and below; even as the
vision in Ezekiel 1.18 which thou beholdest full of eyes before
and behind, above and below, so is the body of Christ; the Holy
Trinity shines in its whole body and needs neither sun nor day."
In the mystical vision right is most certainly not better than left!

A modern Cabbalist, Francis Warrain, lays emphasis on the
correspondence between Boehme's dualism and that of the
Cabbala, in which man is represented as a "movable equilibrium" embodied in the symbol of the scales. The relation
between the two basic principles of the urge to life and the will
for good is of such a kind that neither of the two can exist alone
without regard for the function of the other. Although in the
tree of the Sephirot the termini of the right and left side are like
opposite poles, a steady current goes from one pole to the other.
But the German mystic, so Warrain concludes, has consciously
or unconsciously exchanged right and left on the tree of the
Sephirot. Instead of basing himself upon the side of wisdom,
mercy and victory he has based himself on that of intelligence,
strictness and fame. That is why he came near to the abyss "and
if he did not plunge into it, this was only because a higher
instinct for virtue, a sincere faith and an honest humility
preserved him".[8]

Luther's saying which I quoted just now, *Dextera Dei ubique
est*, seems at first to be very far removed from the dark depths of
neo-gnostic and Christian-Cabbalist doctrine, but there is more
to it than a "judgment of Solomon" in the quarrel concerning
the Lord's Supper, in which the reformers insisted that the body
of their exalted Lord was enthroned on the right hand of God.
The nature of God's creative activity *sub contraria specie* is at issue
here, but also the attitude of the "true believer", for whom "the
great contradictions of existence must in the last resort be
relativized and neutralized".[9] We shall deal with the first of these
two problems in connection with the Leibniz-Clarke debate.

Here I shall only recall the paradox that lies in the existence
of God as an infinitely perfect being. Erich Unger (cf. note 5 to

Chapter I) shows in an unpublished essay *Die Erzählung vom Garten Eden* (the Story of the Garden of Eden) that even in the Biblical narrative this paradox is not fully resolved, that in one sense all that emanates from God is divine and good but that in another sense evil too comes from God, i.e. evil has a function in the world and so has been determined in the will of God. This paradox leads to the creation of free will.

The Manichaeans could see no escape from this paradox save by "drawing God out" of the world and confronting him antithetically with it as the Stranger, the utterly detached one; they gave the Devil half his kingdom, i.e. they "let him fiddle with the left hand", thus making a breach in divine omnipotence. So a vast tension arises between good and evil, a tension which is symbolized by the right-left polarity. Unger tries to rescue the omnipotence of God by means of a subtle dialectic and to show how in the end all things are determined in the will of God. This problem—together with that of grace—forms the very heart of all Christian and Jewish theologies. Thus in a recent book on Jewish mysticism[10] the "first vision" of the Cabbalists is described as follows: the creation of the world through the operation of *Zimzum*, through which God, who is the infinite voluntarily withdraws from a part of Himself in order to leave room for the finite universe of death and time. In all such cases God, one might say, must surrender a part of His symmetry "for the sake of the multiplication and exaltation of reality", so that the creation may take place.

Closely related to this is the idea that the creation is incomplete and that God needs man to complete it. Among Christian mystics this idea is expressed with touching simplicity by Angelus Silesius as follows:

Ich weiss, dass ohne mich Gott ein nicht nu kann leben:
Werd ich zunicht; er muss *von Noth* den Geist aufgeben.

("I know that without me God cannot for a moment live; If I am gone, then must he expire.")

III

THE EXPERIENCE OF RIGHT AND LEFT

1. *Is there a right-left sensation?*

Do we know what is right and what is left? Hitherto it has been tacitly assumed that we do. Now let us consider the difference between left and right as a problem. Let us ask ourselves whether there really is such a difference.

According to one widespread view, right and left have the same existential status as red and green, that is to say that of a sensation, and there is such a thing as a right-left sensation, or a right-left sense. In order to prove this, a psychologist, Kurt Elze,[1] resorted to a well tried scientific tactic which consists in making the nature of a phenomenon apparent by observing the changes and disturbances to which it becomes subject. The discovery that some people cannot immediately tell the difference between right and left led Elze to coin the phrase "Right-left-blindness" which suggests a parallelism between the ability to distinguish between left and right and the ability to distingish between colours. He sees in the ability to distinguish between left and right a *primary independent function of the brain*, comparable to the absolute sense for tones. This power of distinction—or its absence—has, he declares, nothing to do with intelligence. Animals such as rats and hedgehogs can easily be trained to

53

make such distinction,[2] whereas it is just the people possessed of rather special intelligence who tend to complain that they cannot make it without some difficulty. Thus university professors and distinguished physicians had confessed to him that they always had first to stop and think in order to recall which was their right side. They then identified it as "the hand that I write with" or, in a particularly curious case, "the side on which my aunt sat at a children's party". Helmholz, he decleares, was "right-left blind" in a very high degree, while Schiller wrote to Körner: "During the whole of the evening I could not make out which was right and which was left" (cf. p. 134).

There are a number of possible objections to Elze's theory. The concept "red" covers a region that has no sharply defined boundaries and which even within its own confines is not homogeneous. There are an infinite number of shades of red. But there are no transitional shades between left and right nor can left and right mingle with one another. Their sharply disjunctive character—the table is *either* on my right *or* my left—tends rather to suggest an intellectual operation, which for some reason or other the person who is "right-left blind" finds difficult.

That animals can be trained to distinguish left and right does not contradict this. The rat in the labyrinth, the hedgehog trained to push an object, may well act "correctly" without really making the distinction between right and left at all, indeed without any kind of "insight" into the situation. Above all, Elze's right-left sense reminds one of the famous "sense of dimension" concerning which Henri Poincaré (cf. p. 64) made observations, in his *Science and Hypothesis*, which are as profound as they are amusing, and which gain a peculiar charm from the relationship between right-left symmetry and the number of dimensions. Poincaré rightly holds that our sense organs "are designed to inform us of *changes* taking place in the outer world. It is difficult, he argues to understand why the Creator should have given us organs that are designed continually to cry out to us "Rember that space has three dimensions!" for the number of these dimensions is not subject to change. Although the right-left sense may be regarded as a "sense without an organ" we are forced to admit that the Creator would hardly have provided us with an "independent function of the brain" whose only purpose was to inform us continually about an unchanging division of the world into left and right.

Another form of "right-left training" described by Elze seems to point to a more satisfactory solution of this problem. "Recruits", he says, "who were so stupid that they could not even distinguish left from right were instructed in their drill in Czarist Russia by having a bundle of straw tied onto the right leg and a bundle of hay onto their left. In this way they acquired the faculty of orienting themselves "by the guidance of their bodies" and of indicating the position of objects to their fellow soldiers (and above all to their sergeant!). Nothing shows us more clearly than the introduction of this device of "hay-leg and straw-leg" into the learning process that the distinction between left and right is wholly devoid of the elementary character peculiar to sense impressions and that from the very start it involves a "higher" activity, namely the act by which man associates something with his own bodily pattern, a pattern of which he is made aware by the so-called kinaesthetic sensations.[3] This act of "associating" is a comparatively "primitive" one— it precedes that of counting, for instance—but it is undoubtedly an intelligent and more or less essentially human act.

It is in harmony with this conception that, as far as we know, people cannot suddenly become "left-right blind" as the result of an accident or of some bodily injury in the way that a person afflicted with blindness can no longer distinguish colours. More important still, there is the fact that there are no *completely* right-left blind people, so that it would be senseless to say that we could no more explain to a right-left-blind person the difference between right and left than we could explain the difference between red and green to one who is colour-blind.

For Ernst Mach,[4] colours, tones, times, spaces, all these are sensations, and left and right have the same elementary character. "Physical space"—so Richard von Mises summarizes his ideas—"is differentiated by the sensations of above and below just as it is by those of right and left. It has definite directions, which are different in character from each other", directions, one might add, which are apparent to animals too. Mach's attempt, which he makes in a number of his writings, to apply his theory of the primacy of sensation to the "difference" between right and left is in many ways illuminating and reveals him as the first victim of right-left mysticism.

Following the widely accepted principle of psycho-physical parallelism, he seeks to establish that every peculiarity pertaining to our space sensations must have some corresponding physiological peculiarity; this is the principle of the eye as a camera, of the ear as a resonator. Similarly above and below, far and near, right and left, all being different sensational qualities, must in each case have their physiological counterparts. Yet, try as he will, he can find nothing in our visual apparatus that could make possible the distinction between right and left in this symmetrical organ. He therefore indulges in guess work. Are we here concerned with differences in the movement of the eye muscles? With an asymmetry that escapes our notice? Perhaps with chemical rather than organic differences? In the end he considers the most probable explanation to be that the "better development of the right-sided motor function might lead to a modification of the sensations connected with it . . . but a confusion between left and right still occurs in regard to shapes that have a purely optical interest and no motor interest. And in such cases as these the various sensations are 'deafened' by similar, stronger ones. The symmetrical motor organization, which we have acquired in the interests of speedy locomotion, has the effect of making both halves of a spatially symmetrical form *appear* equivalent to us."

These are not merely the unavoidable paradoxes of empiricism. Unless he had previously made up his mind that right and left are "really" different, Mach would hardly have come to regard his sensory impressions as sensory delusions. There is a particularly marked element of paradox in the reasoning expounded in one of his popular lectures entitled *On Symmetry*.[5] He is disturbed by the fact that a child, when it learns to write, is more likely to confuse a and d than b and p; he avoids the obvious explanation which is to be sought in the bilateral symmetry of the human body and in the writing process itself but again seeks a reason for it in the visual apparatus. "The retinas are in fact like my two hands. They too have thumbs of a kind, though by the thousand, and index fingers, though again by the thousand. We can regard the thumbs as being directed towards the nose and the other fingers as being directed towards the outside of the face. When we regard a T, that is to say when we feel its shape with these optical hands of ours, then the right half of this T produces the same sensation in the left half of the retina as its left half produces in the right half of the retina. The T in its entirety thus, like every symmetrical figure, produces the same sensation twice over."

For Mach this repetition also explains the pleasant impression that symmetry makes upon us. But he finds it puzzling that people born one-eyed should have a certain feeling for symmetry and should derive pleasure from it.

Behind this whole way of looking at the matter there lies once again the tacit assumption that the right and left halves of the T do "in actual fact" evoke different sensations. This idea strikes the modern reader as definitely atomistic and un-*gestalt*-like. Even though all psychologists may not subscribe to it, *gestalt* psychology, supported as it has been by developments in cybernetics, has unquestionably bequeathed a certain style to all present-day psychological research, a style which henceforth rules out certain ways of thinking. The *Sehfeld* (field of vision) consciously perceived—as distinct from the merely perceivable *Gesichtsfeld* (area of view)—cannot be a half and still less a right or a left half. There is no perception of half a T nor any other way of experiencing it. A T is not made up of a right and a left half!

Mach's strange explanation of the difference between right and left is not merely a weak point in his system, in which he is actually endeavouring to prepare the way for the substitution of function for substance in the sense of recent logical positivism. It also ill accords with the quite modern ideas which he expresses elsewhere—particularly in his *Mechanics*. What ancestral mythical ideas caused him to become as it were untrue to himself?

In his *Analyse der Empfindungen* (Analysis of Sensations), Mach declares that throughout his life he has been haunted by the right-left problem. "The thought that the distinction between right and left is based on an asymmetry and possibly, in the final analysis, on a chemical difference, has pursued me since my youth." In this connection he mentions how impressed he was by the story of an old officer who told him that troops which had lost their way in a snow-storm and had lost all means of finding their bearings, instead of going forward in a straight line, moved right back almost to their point of departure. What really fascinates Mach, and what he vainly seeks to build into his theory of knowledge and into his psychology, is the "polarity" of the physical world, the existence of a physical space with "its three directions, all different in character, which in a triclinic medium show up their differences most clearly in the behaviour of an electromagnetic element. These physical characteristics", he remarks, "are also apparent in our body, which is why the latter can be used as a reagent."

In his book Weyl describes the dilemma of young Mach when faced with Oersted's experiment (cf. p. 165). But we must not conclude from this that in his later years Mach recognized and analysed his mistake in the way that Weyl was able to do. Even in the last essay of his *Populärwissenschaftliche Vorlesungen* (Popular Science Lectures) which he wrote in extreme old age, Mach clings to the mental experiment of the swimmer who puts himself into the positive electric current as though it were the Thames and discovers to his surprise that the north pole of a magnetic needle points to the left side of his body. And he still asks himself the vain question: "What has our body to do with current or needle?"[6]

The body as a reagent! Is this not obviously the lovely but

dangerous old doctrine of macrocosm and microcosm? Mach is by no means the first scientist to be seduced by it, together with (or, better still, in conjunction with) polarity. And yet, for all his false examples he seems to suggest a perfectly sound idea (a frequent occurrence in the history of science), or at any rate one now recognized by all modern scientists—namely that experiments in physics can disclose hitherto unsuspected properties of space (cf. p. 163). Mach has not expressed himself clearly about this. He suspects, however, that nature in a certain sense is asymmetrical, i.e. "that it is after all in some respects a one-sided individual, whose counterpart does not exist or is at least unknown to us".

In those days people did not yet speak of parity and anti-matter (cf. Chap VI). Even so far-seeing a man, even a man with a true "relativity sense" like Mach, actually denied the existence of atoms. This being so, the words here quoted may indeed be regarded as prophetic.

The two examples discussed show the weakness of the more old-fashioned type of positivism which, apart from all the other objections that have been and might still be raised against it, comes to grief over the problem of left and right. Right and left, or let us say the difference between right and left, are not *sensations*. We must grant them the status of *perceptions* at the very least. This is in line with the ideas of almost all modern psychologists according to whom we are never really concerned with simple sensation, but with perception, that is to say with a complex variable construction in which the kernel of sensation is enlarged by "animal inferences" and is surrounded, as it were, by an aura of memories, volitions etc. As perceptions, or perceivables, right and left should be regarded as something less "elementary" than "red" or others of Locke's "secondary qualities", but for that very reason they should be considered all the more *true*. They would seem to have more to do with the "public" world, perhaps even with the scientific world, which we are inclined to regard today, if not as the true and real one, then certainly as the truest and *most* real.

The epistemologists describe how science gradually does away with the once colourful phenomenal world with its sounds and tangible objects and turns it into an abstract, colourless, but for that very reason less deceptive world, a world which has to be accepted by every thinking person. Only a very small part of the world of our ordinary experience is spared this process of *dequalification* and we shall see that right and left cannot be included in that small part. As yet, however, we have not managed to rid ourselves of them. They are not among the very simple and most directly experienced qualities which people concerned with communicability must make particular haste to eliminate. When "red" has disappeared from the world, a period of grace will still be accorded to the duality of right and left. We must take them along—even if not into the "public world", then at least into its antechamber, into the world of perception.

2. *Space and its problems*

In the realm of perception we come up against space for the first time. It is one of the most widespread views, and one that is hardly ever called in question, that the space of our perception, perceptual space, has a right and a left. The meaning and importance of this observation is not easy to determine, for though in the final analysis there is only one world, there seem to be an inconvenient number of "spaces". The existence of a mathematical space—long regarded as the "space of pure perception"—of a physical space, of a space of action, a space of experience, of an affective emotional space, indeed of an ever greater number of psychological spaces—psychology endows every sense with a special space of its own—the existence, I say, of this wide range of spaces presents us directly with certain problems, for in the case of all these spaces we can at least *speak* of left and right.

We need not be philosophers to recognize that what we have called perceptual space is not a space actually perceived, but must of necessity be a construction from a number of kinds of perception that we call spatial. Before psychology was ever called psychology, it concerned itself with the question: from what perceptions is perceptual space built up and by what senses are these perceptions communicated? Condillac believed that a single sense, which we could choose at will, would be sufficient, while for Berkeley the sense of touch was the only right and proper one. We know today that perceptual space is actually only there intersensorially and that the joint operation of several of the sense organs which furnish our spatial perceptions is required in order to build it up. The over-riding importance of our sense of touch, of which Berkeley was aware, has admittedly been confirmed; it is clear that the unaided sensations of the retina could never produce that experience of a third dimension which enables us to tell right from left and which causes the knots in our world to "hold" instead of unravelling themselves as on a conjuror's stage.

An astonishingly simple experiment shows us that this relation between the senses of touch and sight is what really does govern our ability to distinguish left and right, yet at the same time discloses the precarious nature of the whole distinction. A person standing with his eyes closed, and holding two balls, one heavier than the other, in his hands is asked to cross his arms. Immediately the subject ceases to be able to say with certainty whether the left- or right-hand ball is the heavier of the two. If on the other hand the subject merely closes his eyes, the ability to distinguish right and left is in no way impaired.[7]

The co-ordination of the senses of touch and sight which normally determines our right-left orientation can also be suspended, as Mach has shown, by means of prisma spectacles which interchange right and left. The subject who wears such spectacles lives for the first few days in a fantastically strange world; what was concave is convex, what was distant is near. Then a partial rehabituation takes place and there follows, to some extent at the cost of the optical sensation of right and left, a more or less adequate adjustment to the new situation. Our feeling for right and left thus shows itself to be weaker than the co-ordination between touch and sight. H. C. van der Meer gives us an interpretation of this experiment[8] with the "right-left" spectacles, an experiment which was originally attempted by the behaviourists and later repeated by Kohler. The experiment, which was actually only made to determine the relative importance of the optical and tactile elements, yielded results—and very interesting ones—which went beyond what had been expected of it. To effect a speedy rehabituation a *markedly tactile situation* is needed, by means of which the subject is brought into a new "association of action" with his surroundings. "Only through the manipulation of objects does that which is seen at that moment acquire its direction." Over and above the mere passive sense of touch, the mere tactile stimulation, we can first of all observe the part played by *movements*, we can further note the importance which the "interest" in the action that is being carried out may possess for right-left orientation. Observations of persons suffering from a physical defect are often directed towards the same sort of goal as these experiments, which are directly concerned with the experience of right and left. "Can a blind man who gains his sight as the result of an operation distinguish between balls and cubes which till then had only been known to him through touch?" This was the question that Molyneux had put to Locke. It evoked heated discussion. Berkeley, Voltaire, and Condillac each answered it in his own way. Diderot, for whom (taken together with his *Letter on the Blind*) it earned the Bastille, took the matter up and later began his study of the case of Saunderson who had been blind from birth. The fact that the latter taught mathematics, optics and astronomy in Cambridge shattered his belief that blind people must necessarily have a different philosophy from that of the sighted. Today, though with some reservations, Molyneux's question is answered in the affirmative, but stress is laid on the fact that after the operation the directional control of movements in visual space appears to be restored completely and instantaneously and in a higher degree than the capacity to distinguish between shapes.

But is it really our sense of touch which produces a perceptual space containing shapes and directions? Even Leibniz had already declared that people born without arms or legs, people that is to say who for all practical purposes have no sense of touch, are not in any essential matter different from others. Schopenhauer, too, by adducing the example of a "dysmelic" child, refuted the view that such a person could not "attain to a conception of causality and consequently could not achieve a normal perception of the outer world".[9] I do not know whether any study has been made of the orientation of perceptual space in the case of the Thalidomide victims. Such a study would no doubt confirm the dominant part played by intellectual factors in producing our perceptions.

In the eighteenth century Condillac, the French philosopher, conceived the idea of a statue which at first would only be able to experience a few sensations and would gradually come more to life and become more humanized, until, passing through the various stages of perception, attention, etc. it would attain the highest forms of thought, feeling and willing, and would, in short, develop into an intelligent free person—like its creator, Condillac himself. For a long time the principle of Condillac's seductive "experiment in thought" guided psychologists and epistemologists in their attempts to reconstruct the space of human perception and bring it into connection with other "spaces". They overlooked the fact that what is misleadingly labelled a construction or synthesis was in this case actually an analysis. The starting point of this "metaphysical anatomy"— that is the expression which Condillac himself employs—is not the statue but man fully developed. The psychologist's illusion is that he believes himself able to follow the road leading from the basic elements to himself as easily as the philosopher follows that leading in the opposite direction with himself as the starting point. Kant, being a more honest and logical Condillac, stands

Condillac's metaphysical anatomy on its head by turning the "before" of the psychological construction into an *a priori* of all possible experience; in particular, he makes time and space the formal condition of the latter and regards them as the antecedents of all that is purely empirical.

Kant thus opens the way to critical philosophy, but closes that of psychological research. Actually he was convinced that psychology could be made the subject neither of experiment nor of mathematical treatment and could never become a science. Critical philosophy thus came into more or less open conflict with experimental psychology, which meanwhile had seen the light of day in Wundt's "psycho-physiological laboratory" in Leipzig. This new and rapidly developing science could not, as it were, rid itself of Kantian space. Even Mach the anti-Kantian had to admit: "We can never really get rid of time and space. They are already present everywhere when we observe things outside us or things that are on us or within us."

This then is the form in which the problem of space tends to pose itself, the problem that filled the entire second half of the nineteenth century: how to bring perceptual space into harmony with the "space of the geometricians" as critical philosophy conceived of it, i.e. to construct, by means of things which can be physically experienced, a space which would serve as a suitable frame for our feelings and mental images and which could at the same time be fitted into the space of the geometricians. Our conception of right-left experience will depend on the solution of this problem, for it is claimed that our perceptual space possesses definite directions, whereas homogeneity and isotropy[10] are treated as characteristic qualities of mathematical space.

This problem of space is placed before us with exceptional clarity in the scientific works of Henri Poincaré,[11] which have found a very wide circle of readers in many countries. But Poincaré was the last person to whom this problem presented itself or could ever have presented itself in this particular form; the star of this great mathematician and physicist began to wane when Euclidean geometry finally lost its dominant position.

Poincaré's great achievement is to have applied systematically to the space problem the *group concept*, which is already implicit in the work of Helmholtz,[12] and to the development of which Poincaré himself made a substantial contribution. This enabled him to undertake a scientifically exact description of the actions and changes in position of the whole body (taken in their totality) which contribute to the construction of space; in conjunction with this he was able to "modernize" the old scheme of the statue. For if the statue is capable of "learning geometry", the reason for this, according to Poincaré, is that our perceptual space is in a precise mathematical sense very *like* geometrical space. We are of course unable, he declares, to "put" sensible objects into a geometric space, but there is a certain advantage in being able to show that the two spaces are approximately alike in structure. The building-up of our perceptual space (representational space) on the basis of the co-ordination of bodily movements and of the possibility of distinguishing changes in position and situation, the confrontation of this space with the Euclidean one—these, broadly speaking, are the two themes of his perceptive synthesis in *Science and Hypothesis*, a book still well worth reading today. Did he successfully carry out his intentions? Not altogether.

Between the space of experience by which we are confronted and the space of the geometricians as the scientist works it out and uses it, Poincaré can no more establish a satisfactory relation than anybody else (we shall see later how to achieve this: the two spaces would have to construct themselves "one upon the other"). Thus the sensible data provide a physical continuum with six dimensions, while the geometrician only requires three for his Euclidean space—a fact that is also of interest to us in so far as there are profound differences between the spaces with odd and those with even numbers of dimensions. Above all, however, there is the still unanswered question of why our experiences—both those of ordinary life and, according to Poincaré, those of science—lead to a Euclidean continuum, despite the fact that "the general idea of the group pre-exists, at least potentially, in our minds". Non-Euclidean geometries[13] which have already been known for a considerable time, are now being taken seriously—thanks not least to Poincaré himself. Their equality in mathematical value suggests that all possible geometries, including the Euclidean, should be treated as equally abstract creations of the human mind. Thanks to the general group concept, that mind is now endowed with unlimited creative power and can bring forth countless spaces from out of itself. Yet this magnificent gift remains unused by Poincaré, who clings to the primacy of Euclidean space, i.e. to the relatively "poor" group of changes of position. For him, the general group concept plays an increasingly equivocal part. It becomes a more and more rigid frame, almost a Kantian *a priori*, and in his *Last Thoughts* Poincaré finally asks himself whether it is not a kind of intuition. On the other hand—and it was this that earned Poincaré, who had once been charged with Kantianism, the reproach of Conventionalism—the tool has changed into a mere toy. In a work entitled *The Value of Science*, which appeared much later than his *Science and Hypothesis*, Poincaré sets a much lower value on his earlier synthesis. He begins to doubt whether things had actually developed along the lines he had described and whether indeed the concept of space "had been formed in the human mind in this roundabout fashion".

3. *The statue and the child*

Poincaré died at the age of 58. Had his early death not prevented him from reading the works of Jean Piaget,[14] he would have learned to his astonishment that as a small child, long before he had started to specialize at the age of 8 or 9 in the group of movements of rigid bodies, he had already acquired a most admirable knowledge of topological mathematics and was by no means a stranger to the general group concept. Fate ordained otherwise. And so Poincaré, with his equivocal theory of space, remained merely the distinguished forerunner—just as in the history of physics his name survives as one who "almost" discovered the theory of relativity. Piaget is not the first who, recognizing the fruitlessness of all "metaphysical anatomy", endeavoured to answer the question which Poincaré must have secretly put to himself: what actually happens, what do we establish when we observe a child instead of a statue? Yet the very title of his principal work, which he wrote in collaboration with Bärbel Infelder, *Epistémologie Génétique*, shows that he aimed far beyond ordinary "child psychology". His developmental psychology claims to do much more than solve the problems of child psychology; it also makes a genuine contribution to the problem of epistemology and tells us how knowledge in its most general sense comes into being. Like Poincaré, he leans in this on the group concept, but meanwhile this concept, thanks to the triumphs of theoretical physics, has gone from strength to strength and has proved its objectivizing power, its ability to create reality, qualities which Poincaré had failed to discern. For that operational realism to which Piaget was committed the group becomes the "meeting point of world and thought". The general group concept does not pre-exist in our minds but shapes itself, simultaneously with experience, whose universality and objectivity it guarantees.

The experimental examinations of children which have been undertaken on the strength of this working hypothesis have completely and irretrievably upset the pattern of our hypothetical statue. Thus it has been found that the constancy of objects which enables us to recognize them again and which, according to Poincaré and Helmholtz, is an essential factor in the construction of our space, is not the origin but the *result* of our sensory motor behaviour, and at the same time of that characteristic property of the group, reversibility. It is something with which the little child is quite unfamiliar a fact unknown to the majority of mothers. If a seven- or eight-months-old child is handed its bottle the wrong way up, it never thinks of turning it round, it simply does not recognize it at all. It has not yet had an opportunity to explore the group of movements which define the rigid body as an invariant. Thus it only knows changes of state not rigid objects moving in space. Very much later—and here again because certain transformations prove to be reversible —the child contrives to identify matter mass and volume. And in general, according to Piaget, it is much less the function of reason to identify than to construct and compose.

Today, when the real nature of a child's development is so much better known, there can no longer be any support for the naïve view that the new-born child opens its eyes on a ready-made world which is spread out before it, on objects in a space which possesses certain properties. Rather do we ask ourselves at what point of time in the child's progressive conquest of the world does this or that factor make its appearance. Only observation and experiment can answer these questions, and the answers they give are often surprising. The double ordering CBA-ABC of right-left symmetry appears at an early stage, but it appears in the realm of time and not of space. Before the child can distinguish right from left, it learns the importance of putting on its vest before its jacket instead of vice-versa, and realizes that the order in which it does these things has to be reversed when it undresses. As to the matter of space, the child's "active space" normally begins with "topological intuitions" long before it becomes simultaneously projective and Euclidean. Before we learn to keep at a distance we recognize concepts, or rather, we operate with concepts, such as closeness, separation, surrounding, order, etc.

The fact that this normally remains concealed from us, and that we regard the "Euclidean intuitions"—straight, angle, quadrangle, square and circle—as the initial spatial concepts, is due to a very curious phenomenon to which the school of Piaget attaches great importance, an attitude in which it differs from that of the other distinguished French-speaking psychologist, Henri Wallon. The phenomenon is this: everything that the sensory-motor intelligence of the little child constructs is reconstructed again *ab initio* at a later age under the influence of new experiences which work over again what has already been acquired. In particular, the representations produced by cognitive activities turn back to reshuffle, as it were, the perceptions produced by concrete activities at an earlier stage. The adult human being who has lost all recollection of these earlier stages believes that from the very start every act of perception is using a system of co-ordinates or the vertical-horizontal relationships, whereas in reality the construction of this highly complex intellectual framework only came to an end at the age of eight or nine and has been subject to far-reaching influences due to education and social environment. The result is a misunderstanding of the relation between perception and representation, the erroneous belief—which we can even find in Poincaré—that sensory-motor space provides all that is essential in geometrical representation and that the intellect concerns itself with a sensible that is fully worked out. So too, the illusion of a direct "perception" of right and left can be explained by this phenomenon of "feeding back".

When the matter is regarded in this light, there is no longer any difference in principle between the child, which identifies objects by exploring the "group of movements" and at the same time builds up our world—just as Krishna plays with this globe of ours—and the physicist, who "tries out" a group, i.e. sets a scientific object into a pattern of relationships and thus works out a structure which he then submits for experimental confirmation. Euclidean space, with its three "directions", exists neither for the child nor for the researcher. Both live and work in a sort of aquarium world in which there is no "distance" that the eyes, which Schopenhauer[15] called extended hands, can survey.

The younger the child the more does *proximity* outweigh all the other organizational factors which make their appearance in the course of time; and the topological relationships, which are the first to come into being, are, as it were, the last things that remain at the end of the journey to knowledge. The mathematician's definition of space rests wholly on a purified concept of neighbourhood. As distinct from all other kinds of multiplicity, an abstract space is a multiplicity for which neighbourhood relationships can be defined. In the same manner the theoretical physicist gropes, if one may use the term, from one neighbourhood to another, seeking the structural laws which define neighbourhoods for him. According to the field theory viewpoint of relativity "local happenings are to be significantly correlated . . . not with distant happenings but with happenings or conditions in the immediate vicinity, the aggregate of which constitutes the 'field".'[16]

The little child does not yet know left or right, and the scientist has abandoned them! That is the net result of the combined efforts of psychology and epistemology in regard to the right-left problem. If we assume an "ideal pattern" which stretches from the child to the theoretical physicist, we could therefore say that only for a relatively short span of his individual development does the future scientist stay in the pre-scientific stage, in which he distinguishes right from left. While most of us carry this pattern over into the next, the Euclidean stage, the scientist, being more advanced than the average man, becomes "right-left-blind" before he finally loses even the sense for the Euclidean. Perhaps we can see in this a partial explanation of Elze's statement, which was quoted at the beginning of this chapter, that it is precisely the intelligent who find difficulty in distinguishing right from left. My guess in this matter seems less wild if account is taken of the relationships discovered by a distinguished man who was himself "right-left-blind", namely Helmholtz, between right-left-blindness and a bad memory (which often goes hand in hand with it) for "disconnected things" such as stray foreign words, irregular forms of grammar, historical dates—in short all the things with which a scientist, who, like all creative persons, is only interested in the connections between things, does not want to burden his mind but looks up in a reference book, should the occasion require. This does not mean that everyone who is right-left-blind must be regarded as having attained a higher level of knowledge. Nor should another possible explanation be ignored; namely that there could well be a connection between right-left blindness and deficient lateralization (cf. p. 134).

The manner in which these illusions concerning the relevance of perceptual space come into being, the reason why—to quote but one instance—we believe ourselves able to read geometrical properties like directions into it, all this requires further exploration. All modern psychologists stress the influence exerted on our perceptions by habitual acts, by conceptions that have grown definite over the course of time, and by social and speech factors. The social environment in which man is placed from birth "provides him with a ready-made system of signs which his mind makes use of in the building up of perceptual space, it provides new values and an immense number of obligations". Why, asks Piaget, does man try at a particular stage of his development to visualize the spatial conditions instead of simply endeavouring to act upon them? And his answer is "Surely in order to communicate some reality connected with space to *another* or to receive from such *another* just such communication. Apart from such a social relationship there seems to be no reason why pure representational thought should follow after action."[17]

It is no mere chance that Piaget, in connection with his theory of childish egocentrism, deliberately selects the example of the right-left relation to prove his point. Up to its seventh or eighth year the child's visual thought displays "a distorting egoism" since a relationship, once discovered, is referred to the child's own activity as a subject and is not "decentrated" onto an objective system; the child gives precedence, if I may quote an expression of Wallon's, to the optative over the indicative; it is "phenomenon-bound". In this egoistic stage, which is wholly governed by the "centration" process, the child, though it can indicate its right hand, nevertheless makes mistakes about the right-left relations in the case of the partner who confronts it "because it cannot put itself in the position of the other, either socially or geometrically, and ascribes to that other its own way of seeing things". Thus in Piaget's example it is social conduct which first creates the preconditions for the mental image continually suggested to us by our education, the image of a perceptual space with a definite right-left character which exists outside of our own bodies. The right-left organization of our perception takes on the appearance of being nothing more than a pure pattern of arrangement which man develops "with the help of the guiding thread which his body provides".

In these circumstances can we continue to speak of a perceptual space at all? Although modern psychology does all it can to devalue this concept, it has refrained from so far wholly discarding it. "There is such a thing as perceptual space", says Piaget "and everyone must admit this." But he adds the warning that our spatial perceptions, which are not given to us at the very beginning of our mental development but are constructed under the influence of a variety of factors which were originally alien to them, must not be regarded as though they were a sort of disinterested knowledge, let alone the starting point of new and more complex processes of cognition.

4. *On the borders of psychology*

Perceptual space, in which right and left should be directly discernible, began to fade away in the light of a scientifically oriented psychology. It no longer confronts us as a datum characterized by certain specific properties. As in a Spanish inn we only find there what we have ourselves brought in. Nevertheless it remains an object of psychological research, and the question why in a certain case perceptual space is endowed with a characteristic right-left structure can present a genuine problem. (Thus one could, for example, draw valid inferences concerning the character of a child from the changes of direction in handwriting which frequently take place during puberty, by interpreting them as a protest (cf. p. 81).) Assuming, of course, that Piaget's warning is not disregarded and perceptual space objectivized and made into a source of knowledge. There are plenty of instances of this, however.

H. C. van der Meer[18] attributes to the space of our action and perception—which she calls phenomenal space—a number of unchanging discernible qualities, above all an asymmetry or polarization towards the right. "In vital and experienced space there is no similarity and interchangeability of dimensions. Left and right stand here in polar opposition. The main directions divide space into world regions with their peculiar qualities of value; they are significant and effective as sense directions. According to this author's investigations (some sixty persons were questioned!) these sense directions are the *directions of time*. "We can thus regard the directional asymmetries as the manifestation of the way in which our being-directed-towards-the-future is lived and experienced." This experience is implicitly effective both as an incarnated norm and also as a value. By equating right with future and left with past, Van der Meer connects the right-left polarization of "phenomenal space" directly with a whole list of Pythagorean pairs of values. Is not the future "better" than the past? "Since man is an expansive being," concludes the Dutch psychologist, "and so is directed towards movement and activity, we can expect the right side to be the side of freedom and activity and the left to be the passive side, the side of constriction." And elsewhere she writes: "One could assume that there is a preference for orientation towards the right where Western European culture is concerned."

The reader who has not as yet been able to divine such "sense-directions" can only see a confusion of scientific thought in this welter of spatial, temporal, and value definitions, to which we must even add—and this is characteristic—political ones. But this kind of thing is by no means exceptional. We all know that, in addition to its classical nucleus, which is oriented towards natural science, modern psychology covers a very wide field, a field which ranges from the various depth psychologies to the so-called sciences of expression. The methods of research used here are as widely removed from those of classical psychology as are the methods of microphysics from those of classical macrophysics. Their claim to be scientific varies very widely. And it is precisely in these border zones, if we may still give them this name, that the old myths of right and left come to life again, or have remained alive. Perceptual and experiential space frequently merge into a polarized phenomenal space from which people believe they can read out whatever they have read into it. It contains a right and a left, and right is generally better than left.

Like the "idealist morphology" the sciences of expression for the most part have their roots in the intellectual atmosphere of that age of natural philosophy when, as Eckernamn tells us, Goethe and Lavater slept in one bed. The ideal of all sciences of expression is formulated in the hope of which Lavater speaks when he says: "A higher, angel-like intelligence could deduce from a joint or a muscle the whole external formation and the all round contour of the entire man; consequently a single muscle would suffice for us to calculate his whole character." Actually the science of expression is becoming increasingly divorced from the principle of *physiognomics*, which interprets the signs provided by bodily build as though they were some kind of cipher, and is concerning itself increasingly with the play of living movement. According to August Vetter, Klages transformed physiognomy into *movement interpretation*. He understood the picture presented to him not as a matter of shape, but as motion and flux. The sound idea which lies at the bottom of all this is that the interpretation of gestures, which are produced by habit or necessity, and sometimes deliberately made, can be more revealing of the character of a man than the lines of his hand or the shape of his skull.

We cannot of course wholly exclude the possibility that in a science of expression which has consistently divorced itself from the visible form, the choice of directions may possess a certain significance. But a glance at the most "scientific" of the expression sciences, graphology, shows us that the necessary preconditions for anything of this kind have not been—and indeed cannot be —fulfilled. This is just what we can discern in the importance attached to right and left by the graphologist.

The graphologist does not concern himself with the overall direction of the writing, and there is of course no cursive script that goes from right to left apart from the mirror writing carried out with the left hand by a fully left-handed person. Indeed in the interpretation of handwriting, and in that of many projective tests, the handedness of the writer is treated as being of no significance. And strangely enough the graphologist seems unable to tell from the writing whether it is that of a right- or a left-handed person, of an ambidextrous person or of a fully left-handed one, etc. Just as he clearly cannot tell for certain whether a specimen of writing is that of a man or of a woman.

On the other hand, it is a matter of great importance, from the graphological point of view, whether writing slopes to the right or the left, whether the strokes of a T, instead of being straight, slope in one direction or the other, whether the O's are open toward the left and whether the margin on the paper is wider on the right than on the left. All this is regarded as significant. And here again we encounter a space with essentially different directions, for the peculiarities of handwriting do not simply mean that the gesture in question can be carried out more easily in one way rather than in the other because of some inborn factor or even because of a special acquired skill. *The directions of space themselves have a definite significance*. For instance the left half of the typing paper that lies before me has something to do with "inwardness" and the right with "outwardness". This is no humorous exaggeration but the almost inevitable result of the way in which graphology has been built up into a communicable science. The attempt to systematize graphological knowledge, to set up rules for the interpretation of handwriting and perhaps compile dictionaries of symbols, ultimately leads us to objectivize an oriented writing space in which right and left possess a particuler "intrinsic" significance.

And so, from the first graphological efforts of antiquity down to the modern school of symbolists who deliberately base themselves on a mythical-archaic pattern of the world (Max Pulver et al.) every graphological system has its "tree of Sephirot". Nowadays, of course, the left is not simply designated as evil; the two dominant categories are introversion (left) and Extroversion[19] (right)—two concepts which clearly involve value judgments: for the extrovert, the person who looks outwards towards the real world, is the social, useful person, the person "who is easy to get on with"; introversion on the other hand only has positive significance in the case of the artist and even there has certain overtones of perversion and the pathological. On the modern graphologist's "tree of Sephirot" the left-hand branches tend to be called egoism, cupidity, base sexuality, etc. and the right-hand ones altruism, goodness, an urge to give and joy in creation.

This general pattern—we must emphasize—is subject to a number of refinements in its application. In particular we are urged to regard every sign, every peculiarity within the context of the whole. But this whole is in actual practice usually nothing but the picture presented by the writing in its totality. The right-left duality as such, moreover, is subordinated to other dualities, possibly to a general division into "higher" and "lower" writings. For instance in Ludwig Klages[20] we find a superimposed classification, in which the writings are divided according to their "Formniveau" (the degree to which the form of the writing expresses the writer's personality) into positive and negative. A characteristic in a positive handwriting has not the same significance as it has in a negative one. Thus inwardness, which goes with the tendency to run leftward, appears in a positive context as contemplativeness and in a negative one as egoism. In Table XIII in his book *Handschrift und Charaktür* the heading *Linkslaüfigkeit*, "tending to run to the left" (urge to appropriate) contains the positive sequence: energy, self-reliance, power of decision, power of survival, business sense, while the negative sequence runs: selfishness, envy, lack of feeling for others, egoism, greed, malice and vindictiveness. We make the strange discovery that on the right side there are twice as many positive qualities as there are negative and on the left twice as many negative as there are positive.

With Klages we encounter for the first time the concept of the symbol in the analysis of handwriting, though he does not attach nearly the same importance to it as the school of the symbolists and especially Max Pulver.[21] For the Swiss graphologist two equations are basic: $I = present = left$; $thou = future = right$. These derive from the fact that we write from left to right "away from ourselves and towards the other person". This direction of our writing is embellished by him with all manner of positive values; speaking generally our manner of writing "symbolizes extroversion, the expressing of oneself. My communication moves from my 'I' to a 'thou'; the writing road is the bridge which is built from the 'I moment' forward into time, i.e. towards the future." But the future does not lie only on the right: "indirectly connected with time-symbolism there is yet another that underlies the latter: the tension of 'past-present-future' reposes, psychologically speaking, on the bipolarity 'mother-father'. Under time-symbolism, with its tendency to run from left to right, we become aware of the build-up of the system of ideas: We derive from the mother and strive towards the father, from physical bondage to spiritual freedom and power. Mother and past are synonyms, so far as our symbolic experience is concerned, as are father, future and fulfilment." Hence the affinity with the mother-world of all the strokes that have been widened from or to the left, while all that widen towards the right have an affinity with the father-world.

On the basis of his topography of experiential space Pulver develops a national—or racial—psychology. That Semitic peoples write from right to left shows him—for he fails to distinguish between cursive scripts and the drawing or painting of individual letters—that they are reflective and turned towards the world within themselves. "The script that runs towards the left transfers its values outside the historical process and back to the origins. The past contains the most essential realities, the patriarchs are more important than their descendants, all decisive events took place long ago, it is the old paradise that they are seeking to regain, not the new Jerusalem. In the movement towards the right we can trace the gestures of Chiliasm from out of which European history came into being and by which it is sustained." After such samples as these, it will hardly surprise us if in other more primitive sciences of expression the right-left polarity of space should be credited with exercising even greater influence. In an interpretation of the *hand* by Charlotte Wolff[22] we again find the two lists of the good right and the evil left and this time they are longer than ever. Moreover in this science of *chirognomics* the direct connection with left-handedness, a quality at which some people already look askance, still further intensifies the polar tension; indeed the right hand is here described as being "more intelligent"; it contains the organizing framework not only of all altruistic feelings but also of the function of judgment.

There can hardly be a single physiognomist who has explicitly attributed any significance to the two halves of the face other than that due to functional asymmetry. Nevertheless let me here tell of a rather extraordinary experimental technique and one which on closer examination seems distinctly equivocal; it created something of a sensation at a congress of the *International Institute of Anthropology*. By sticking together two right halves and two left halves of the same photograph symmetrical faces were produced of a kind that never occur in nature. These the subjects in question were asked to examine. We are told that the left-left faces were appraised quite differently from the right-right ones. The remarks made about the former referred to the external life, the "public relations", the social personality, the gathering of external impressions; the remarks about the latter however concerned the inner life, the personality in depth, the assimilation of impressions received. In a certain French encyclopedia an entire page has been devoted to these strange faces and this equally strange theory. The author seemed worried by the fact that the verdicts were exactly opposite to those which might normally have been expected from people engaged in this particular science. I believe that the whole matter can be explained quite simply: the round jovial left-left faces really do suggest social contacts, while the long, ascetic emaciated right-right faces call up the image of an inward spiritualized life. That the two halves of our faces are different, however, is something that the accepted theories of physiology are probably quite capable of explaining.

Whether this technique has given rise to a "projective test," I do not know, but it lies half-way on the road to this widely used experimental technique of psycho-diagnosis.

In the *projective test* the subject is given some material to "reshape", to which he gives form by "describing, selecting, drawing or manipulating". By this means the subject "projects" outwards his inward self and so reveals his real personality and submits it to the interpretation, that is to say the "calculations" of the diagnostician. Among the various projective tests (Rorschach, Warteg-Test, World-Test, Szeno-Test, the scribbling test of Paul Meurisse, the Colour-pyramid-Test etc.) a certain number are concerned with the structure of space. It is only when we are speaking of these that there is any point in asking whether right and left have any significance or, to be more precise, whether the examiner can know what they signify.

It is noteworthy that the oldest of all projective tests, the Rorschach test, gets rid of the right-left symbolism by means of a "symmetrization". As long as Rorschach merely asked his subjects to interpret simple ink blots, he obtained no results that were of any value. It was only when the papers were folded together and symmetrical figures were thus obtained that the test was of any use. But such precautions are not always taken. Instead, the right-left symbolism which manifests itself in the activity of the subject, is treated as directly significant. In the *tree test* developed by Karl Koch the problematical aspect of the whole projective procedure is made particularly clear by a direct analogy with graphology. In this test the subject is invited to draw a tree. The manner in which he carries out the drawing, and the place where he puts the tree on the paper, shows his personal experience of space. The drawing corresponds to the objective world but has an inner affinity with the spatial pattern of his soul. Koch speaks of the emphasis being on the right if a certain ratio is exceeded between the left half of the tree top and the right half, which is normally the broader one; he also examines the length of the shadows, the richer and poorer forms of the shoots and of the tree itself; like the graphologist he attaches no significance to the inclination of the tree. The right-left symbolism with which we are already familiar from graphology provides the key to his problem of interpretation. A thrusting out to the right is regarded as a "march forward" and sometimes as flight. "Many trees which swell out to the right are like men with rounded chests or protruding bellies. This forward movement expresses the wish for adventures and a desire to be reckoned important." Conversely, left emphasis leads us "into the innermost parts of the ego, and when it is exaggerated it is to some extent as though the author of the drawing was taking cover behind himself. At any rate he draws in his head. From a tendency to introversion it is only a short step to narcissism; sometimes a symbolic element also intrudes: a Russian refugee leaves out the entire right side of the tree; he had had his right leg amputated and had also had to leave his "right hand" behind namely his wife and children. In the case of divorced persons we have more than once come across cases where the right side has been left completely empty. This looks like a repression, a detaching, an extinguishing of oneself, but such a view is utterly mistaken, for the "you" is clearly recognizable as a phantom in empty space.[23]

It is noteworthy that the author of those lines does not run away from questions which threaten to reduce his space symbolisms *ad absurdum*. How is it that in the case of a number of drawings, all made by the same person, many details are sometimes on the right and sometimes on the left, that the shadows so easily change their places, and that a slope towards the right

becomes a slope in the opposite direction? Here Koch falls back
on the obvious scientific explanation; it is that mental instability,
such as occurs during the time of puberty for instance, may
bring a tendency to confuse left and right (cf. p. 72). He also
asks himself the question whether circumstances could not change
completely for the person who uses his left hand to draw, but
answers it in the negative. Owing to lack of material, he says, he
has never been able to prove that the drawings of left-handed
people are mirror images of those of right-handed people (cf. p.
86). Handedness in his view is not connected with the fact that
a person is a "right-personality" or a "left-personality". "Many
right-handed persons move for the most part in left mental
regions and vice versa" (cf. p. 131). Finally Koch recognizes
the relative and essentially social character of our representation
of space and asks himself "whether in the course of time, and
under the influence of the new pictures of the world which
science may yet draw, we may not be on the threshold of a new
way of visualizing space". Nevertheless he clings to his tree test.
"Spatial factors are certainly relative things, but they are not so
indeterminate as to be completely useless as instruments of
orientation in the psychic landscape."

In recent years the method of projective tests has been
violently criticized; at a symposium which took place under the
chairmanship of H. J. Eysenck it was agreed that these tests were
neither tests nor projective. It is undeniable that their "calculation" presents difficulties: whereas in the case of the other tests
only one of two results can be expected—success or failure—
the endeavour here is to obtain in response to a single stimulus a
maximum of qualitatively different answers whose meanings
have then to be co-ordinated. To these difficulties there must be
added, in the case of "spatial" tests, the ambiguity of the right-
left structuring.

In the various depth psychologies the part played by the
concepts right and left is not easy to determine; they occur there
less frequently than might have been expected. With Adler,
left-handedness is regarded as an "organ-inferiority", and is

thus accounted as one of the inferiorities that have to be compensated for and so calls forth a masculine protest, but it is without symbolic value, just as sexual polarity, which Adler ranks below the will to power in importance, occupies a markedly inferior position in this particular system.

To understand the position taken up by Freudian psychoanalysis, we must remember that for Freud it is not the biological but the psychological facts that matter. Just as "dream work" and not the manifest "dream content" is the relevant thing, so the biological phenomenon of left-handedness in itself is without significance for the psycho-analyst. All that matters is how the person deals with it. Even before he created psycho-analysis, Freud had dealt with the problem of left-handedness from the physiological and clinical point of view. He had collaborated with Rie in writing a clinical study on unilateral cerebral paralysis in children. In another essay, which was also written in his early years, *A Childhood memory of Leonardo da Vinci*, he still shared the view of one who at that time was his friend, Wilhem Fliess, on the interconnection of homosexuality and left-handedness (cf. p. 133). But it is actually the primitive-naturalistic conception of the symbol, with which the later analytical schools, and especially the followers of Jung, charged Freud, which protects classical psycho-analysis from a widly luxuriant space symbolism such as we encountered in the sciences of expression. For to Freud the symbol, being a preliminary step towards the concept, is only a sign for something; the meaning of the symbol is a secondary consequence of sublimation. Actually there is for him no such thing as "genuine" symbolism. The symbol remains what it is and cannot develop that expansive life of its own which could if need be bring the entire universe under its spell and divide it into two parts. The abandonment of Fliess's theory of sex, which was based on a supposed essential qualitative difference between right and left, was therefore an essential precondition for the development of his own doctrine. Freud's letters to Fliess, which only became accessible to the reading public a few years ago, confirm that it was the right-left problem which contributed more than anything else to the estrangement between two men who had been inseparable friends only a few years before. For Freud the end of his friendship with Fliess spelt the abandonment of any attempt to establish a relation between the pair right-left and the pair man-woman. It also marked the birth of actual psycho-analysis as such.

For Jung the function of the symbol is much wider—and it is this more than anything else which sets him apart from Freud. Symbols have a universal content. They are visible or conscious archetypes which all peoples have developed in common in the process of phylogenesis. Given Freud's marked positivist bias and his general orientation towards the natural sciences, it was rather from the followers of Jung that one might have expected a coherent system of spatial symbolism and the kind of picture of the landscape of the soul which would be characteristic of the expression sciences. Yet in Jung's depth psychology right and left do not appear to possess the archetypal character of a genuine symbol. This whole system of thought stands under the sign of four and not under the sign of two—under the sign of the interpenetration of Yin and Yang, of the circle and of the cross, all these being symbols of the union of opposites, of the *coincidentia oppositorum*. The *anima*, is here apportioned to the man the *animus* to the woman and the nocturnal voyage of analysis reaches its end with one or more *mandalas* (cf. note 1, 7), which serve to express the self-regulation of the soul.

5. *Right and left in art*

Art criticism and the psychology of art are concerned with an "artificial" perceptual space that has been created by the artist himself. Here the problem is not the same as in the expression sciences. In a work of art the soul of the artist finds expression. But in a work of art the artist also expresses his soul. The conscious factors tend at least to balance the unconscious ones. That is why a work of art is only partly a "projection" in the sense in which psychology uses the term. Without in any way detracting from the part played in artistic creation by the dream, by inspiration and the irrational in general, we can assume that the artist who puts a figure, a form or a colour on the right or the left side of his picture—as the case may be—*knows* what he is doing. The question therefore by which we are confronted is this: do right and left each have a different significance for the *viewer*, of whom the artist is thinking, or as properties of space? According to Heinrich Wölfflin[24] the right and left sides of a picture have by their very nature different "mood-values". Hence pictures cannot be looked at in their mirror image without producing a totally different effect. This is a fact to which the great Swiss art-historian attached very great importance, and this led him to regard as a matter of some moment the circumstance that as a result of reproduction techniques—apart from photography—the sides are interchanged. The discovery that many distinguished artists have nevertheless often put reproductions of their pictures on the market with their sides reversed will therefore occasion some surprise. There are several such reversed versions of Rembrandt's *Christ Taken Down from the Cross* which hangs in Munich. It is said that as time went on the artist became more sensitive in this matter. In this connection we are bound to ask to what extent preliminary sketches and designs can be regarded as works of art claiming our attention in their own right and as endowed with certain specific pictorial qualities of their own. Wölfflin declares that "the decisive element, which is definitely non-reversible, is only added in the last stage of the execution". He substantiates this theory with the aid of two works by Dürer: the drawing, *St. Jerome in his Cell* which is in Milan and was used for a woodcut in 1511, and the drawing, *The Death of Mary* dated 1510 which is in Vienna. Both drawings differ from the prints in that they are reversed.

Mercedes Gnaffron[25] took over the argument from the point

where Wölfflin had left it. Like Wölfflin, she declares that if she looks at two pictures, one of which is the reverse of the other, she can tell which is the "right" one. In her view many preliminary studies for woodcuts, etchings and engravings which show the subject exactly in reverse give a more convincing rendering of their themes than do the prints. The majority of the prints of which no preliminary studies have survived only become artistically convincing when we see them reflected in a mirror, which is how the artist saw them in his mind's eye and set them down upon the wood or metal. To prove her point, Mercedes Gnaffron showed a large number of people some graphic works, both prints and on the plate, and let them decide which in their opinion was the "right" picture. The result left absolutely no room for doubt, though the judgment of experts was less unanimous, prompt and assured than that of the less sophisticated viewers. For instance the print by Rembrandt, *Landscape with Three Trees*, which could be proved to have been reversed, was regarded by Wölfflin as showing the composition that the painter had actually had in mind.

I am compelled to admit that when as an "unsophisticated viewer" I looked at the numerous illustrations in Mercedes Gnaffron's book, which always showed the two enantiomorphic versions, I, like Wölfflin, held the "wrong" representation of the three trees for the "right" one. I do not want to say too much on this difficult subject, but it seems to me that in landscapes there can be no question of a difference in "mood value" but that in the case of figure painting we tend to look for the centre of gravity, if not in the middle, then towards the top right hand. This is particularly true of scenes which are familiar and are often chosen as subjects—religious pictures, for instance. On the other hand it is a matter of comparative indifference to me whether St. Jerome sits on the right or the left of his cell.

A number of inferences have been drawn from these as yet very fragmentary observations. Wölfflin believes—on the strength of the principle that "right is better than left"—that the "good places" are actually to be found on the right of the picture, Moreover, he thinks that we have a tendency to scan a picture from left to right and in a rather special way, raising our gaze at first and then lowering it, thus tracing more or less the figure of a triangle. There is also, he claims Hans Curlis's "law of sight-direction", according to which there is a natural right and left in the picture. Mercedes Gnaffron empirically deduces a "law of sight" which derives from the functional asymmetry of the brain and which brings about an apparent shift of the beholder's viewpoint towards the left half of the picture. She infers from this that a completely left-handed person—Leonardo serves as her example—sees "the sides the wrong way round".

Let us further note that Clouzot's Picasso film gave many people the opportunity of unconsciously repeating Mercedes Gnaffron's experiments. They saw Picasso painting on a sheet of glass and witnessed the creation of a picture in which right and left were in the reverse order to the artist's intentions. It would be interesting to establish to what extent this reversal was noticed and whether people with exceptionally strong right-left sensitivity were upset by it.

IV

SYMMETRY AND DISSYMMETRY IN
ORGANISMS

1. Asymmetry as the mark of the organic

Does anything in the world escape the domination of the principle of symmetry?

Many notable thinkers have discerned in asymmetry a mark of life or, more generally, a mark of the organic. Contrary to the unrelievedly symmetrical physical-chemical world as it hurries to its final dissolution in entropy-death, the biosphere, it is said, is by nature asymmetrical. Much as the artist modifies the strictness of absolute symmetry by some asymmetrical detail— "Nothing oppresses the heart so much as symmetry" says Victor Hugo—nature has furnished the organic world with a certain measure of asymmetry. There is more involved here than the mere fact that the "hexagonal absurdity of the snow crystal"—to use the words of Thomas Mann—is felt to be life-denying; there is also the difficulty, which must be taken seriously even from the point of view of science, of harmonizing with perfect symmetry phenomena such as growth and development that are typical of life.

Aristotle stressed the peculiar position of living things. He assumed, however, that the heavens were supremely alive,

87

supremely *animate* whereas we regard them as lifeless. He gave
them a left and a right as he did the animals. And since this
eminent observer of nature failed to notice that hops wind
around the pole from left to right while the bean winds from
right to left, he made a rather startling differentiation between
animal and plant. For plants, he said, there is an up and a down
but no right and left. Yet it was precisely from plants that
Goethe worked out a concept in which one could see an example
of nature's tendency towards asymmetry. In his work on *The
Spiral Tendency in Vegetation* he declares that the construction of a
plant is governed by two principles—the linear vertical and the
spiral. "In the formation of vegetation the vertically uprising
system produces the permanent element, the element that
solidifies with the course of time, the lasting element, the fibres
in annual plants, a great part of the wood in perennials. The
spiral system is the producer of growth, the multiplier and
nourisher; as such it is a transitory thing and, tends to isolate
itself, as it were, from the other. If it continues to exert an
influence when there is superfluity, it soon weakens and is
subject to decay; joined to the other, both form a lasting unity
in the form of wood or some other solid substance. Neither of the
two systems can be conceived of as existing in isolation; they are
always and eternally together, but in complete equilibrium they
bring about the perfection of vegetation."[1] These remarks end
with a typical reverie on the subject of polarity: "If we regard the
first of these two as definitely male and the other as definitely
female, then we can conceive of all vegetation as being in secret
androgynously connected, whereupon, following the changes of
growth, the two systems draw apart in open opposition to each
other only to unite with each other again in a higher sense."

Needless to say such aberrations of natural philosophy are
today relegated by the majority to the "deepest depths of mysticism". Yet the spiral tendency of plants and the interpretation
of it as part of a general tendency towards asymmetry on the part
of living nature has on several occasions attracted the attention
of botanists and zoologists. Von Haecker, for instance, was

thinking along these lines when he said: "From out of the innermost nature of organic matter the tendency towards asymmetry and spirality seeks to assert itself." To which Wilhelm Ludwig rightly objects that it is as easy to make a spiral line as a straight one or a circle, if not easier; a fish swimming evenly would, if deprived of the use of its sense organs, including its static apparatus, describe a spiral. In his exhaustive enquiry *Über rechts und links im Tierreich*[2] (On Right and Left is the Animal Kingdom) he discusses the possibility that in every manifest asymmetry an aboriginal tendency which had existed before bilateral symmetry had been acquired was forcing its way to the surface. He answers that question with an emphatic "No". Wilhelm Troll reaches a similar conclusion, though in order to leave some meaning in the speculations of Goethe's natural philosophy he wants to see a distinction made between this "spiral tendency" and dissymmetry, which was never primary.

It is only in the work of Louis Pasteur[3] that, despite many mistakes, the idea of a general asymmetry becomes scientifically fruitful. For the nineteenth century's great chemist and biologist *dissymmetry* and not mere asymmetry was the characteristic feature of the organic world. In this term, which was often translated as asymmetry, the whole ambiguity of the right-left relationship can be discerned. It is based on the occurence of pairs of objects each of which is asymmetrical in itself, i.e. possesses no element of symmetry, but which have a relationship of image and mirror-image. In the alphabet the big A is symmetrical, (it possesses a plane of symmetry), the little b is asymmetrical, the pair bd is dissymmetrical if we regard it as a pair, symmetrical if we see in it a single figure. The figure bd however, as we can see at a glance, is less symmetrical than, say, the figure AA. So Pasteur had some justification in regarding dissymmetry as an intermediate stage between symmetry and asymmetry.[4]

Like Mach, Pasteur had his right-left experience. While he was still a pupil at the École Normale, he declared some forty years later, all his ideas had been turned upside down as a result of Biot's discovery that "natural" tartaric acid turned the plane of polarization in the opposite sense from artificially produced para-tartaric acid, and even more by Mitscherlich's declaration that *apart from this optical difference no other difference could be found between the two*. That contradicted the logic of science and Pasteur's achievement actually lay in the discovery of that difference. The crystals from tartaric acid obtained from grapes are all right-dissymmetrical, whereas half the crystals of the chemist's para-tartaric acid are right-dissymmetrical, and half left-dissymmetrical; it is therefore optically inactive, for of the two acids of which it consists one turns the plane of polarization by the same amount to the right as the other does to the left.

There can be few discoveries in the history of science which have had more important consequences. It led Pasteur himself, in the words of Jacques Nicolle—"driven by an inflexible logic, from molecular-chemical research, to ferments and finally to a cure for rabies." [5]

What fascinated Pasteur was the discovery that whereas the known mineral products—whether they were natural or artificial —were never dissymmetrical, molecular dissymmetry seemed to be characteristic of substances produced from living organisms. Sugar, starch, albumen, fibrine, cellulose—all these are optically active. This peculiarity seemed to Pasteur to be connected with their origin. Whenever he had to deal with combinations of a "vital kind", he found that the elementary atoms which formed the molecules arranged themselves dissymmetrically owing to some mysterious influence, and this process seemed to be related to the conditions under which life came into being.

Pasteur suspected that in the final analysis this mysterious influence must be of a cosmic nature. "I believe", he wrote to his laboratory assistant in 1871, "that a dissymmetrical cosmic influence naturally and continuously governs the molecular organization of the principles which are immediately essential for life and that consequently the species of the living world, their structure, their form and the pattern of their tissues are related to the movements of the universe. I should like to establish by experiment a few basic facts about the nature of this great cosmic dissymetric influence. It must, it may perhaps be electricity or magnetism. . . ."

Elsewhere (Proceedings of the *Académie des Sciences*, 1874), he goes even further: "I am convinced that life, as it manifests itself to us, is a function of the dissymmetry of the universe or of the consequences of this dissymmetry. The universe is dissy-metrical, for if one were to place before a mirror all the bodies of which the solar system consists, together with the motions that are peculiar to them, one would have in that mirror an image that could not be made to coincide with reality."

The theory of the cosmic origin of dissymmetry was and sometimes still is used to assert the privileged position of one side, but fortunately Pasteur lets himself be carried away by speculations of that kind as little as Mach; he declares that nature sometimes forms right bodies and sometimes left bodies and for him it is certainly a mere matter of chance that God should have created "right" tartaric acid, leaving it to him (Pasteur) to create, or rather isolate, "left" tartaric acid. Nor was he ignorant of the fact that by means of an ordinary chemical process such as the influence of heat one can pass over from a right substance to a left or vice versa. Incidentally, Pasteur had the opportunity to undertake, or rather to suffer such a right-left reversal in his own person when—after a stroke—he learned at an advanced age to write with his left hand. It is noteworthy that, though his belief in a cosmic influence might well have suggested this, Pasteur never spoke of one side or the other having a preferential status. This shows that, though he was a deeply religious man, he nevertheless "left God outside the laboratory door". In a century in which so many biologists were crippled through having to choose between the barren alternatives of materialism and vitalism, Pasteur had no sooner erected a barrier between the different kingdoms of nature than he began to think of demolishing it. "The distinction which I made in the year 1860, is a *de facto* distinction and not an absolute one or one of principle. *Not only do I believe that this barrier between the organic and mineral kingdoms can be surmounted, I have even indicated the first experimental conditions which in my opinion might be capable of making it vanish.*" "Up to his death", says Jacques Nicolle, "the great scientist passionately sought to change nature by the creation of dissymmetrical bodies."

The ideas of Pasteur concerning the relation between life and "dissymmetry through enantiomorphism" are no longer tenable today. It is easy for us to make fun of the curious asymmetrical laboratory methods by which he desperately sought to produce dissymmetry. He was himself later to call many of them crude, which was basically what they all were. Thus he let salts crystallize out through the agency of a rinsing stream of water passing through a large basin or by means of clockwork, to say nothing of the attempt to get a plant to develop from the seed upward by means of reversed sunlight reflected in a mirror.

Yet how trivial all this is when set against the fact that here was a scientist who by experimental means was boldly seeking to force his way into the intermediate territory that lies between the living and the inanimate. In Pasteur's day the gulf which separated these two was wider than it is today.

The whole chemistry of organic compounds was for all practical purposes unknown at that time. Even more completely unknown were the ultraviruses—formations which from one point of view appear to be living things and in their laboratory or permanent form appear as little white crystals with the same pathogenic properties. Pasteur now stumbled on this intermediate kingdom. By recognizing the relation between the crystalline form of a body, its chemical constitution and the rotation of the plane of polarized light, he laid the first foundation for the theory of asymmetrical carbon and with it for all the structure theories of modern stereochemistry. For a long time this was regarded as a refined metaphysic until at length Max von Laue's remarkable experiments enabled us so to speak to "see" the structures of which Pasteur had been dimly aware. The relation which Pasteur disclosed between optical dissymmetry and dissymmetry of the crystal is still valid today in the majority of cases. Yet there are exceptions, because the dissymmetry of the crystal-forming medium can result—though it need not do so—in the formation of hemiedric surfaces.

Starting out with the idea that dissymmetry is characteristic of life, Pasteur arrives directly at the conclusion that *fermentation*, which even distinguished chemists like Justus von Leibig looked upon as the "work of death", must be a phenomenon of life. This conception has proved exceedingly fruitful, both in theory and practice. We not only owe to it the methods of manufacturing beer and vinegar, it also ushered in the great era of the diastases, ferments and enzymes; the famous lock and key theory by which Emil Fischer explains the specificity of these remarkable catalysts follows directly from Pasteur's right-left observations.

Pasteur showed that in nature there is no preference for a direction, only a dissymmetrical "displacement", as is symbolized by the Chinese myth of Kong-Kong (cf. p. 45). For him right was not better than left. Indeed, regarded by itself it was not even different. But he attached an almost childishly Utopian hope to the consequences that would result for mankind if right substances were substituted for left substances, or vice versa. He was convinced that a new world would open for us if we could turn "right" cellulose into "left", or all blood albumen, which is "left", into "right" albumen, and he visualized the possibility of influencing the processes of nature, to a degree which had never previously been conceived of, by such artificially induced exchanges of right for left and vice versa. "I will not give up my hope of achieving by this means a *profound, wholly unpredictable and extraordinary modification of plant and animal life*." "An achievement along such lines as these would open up a new world of organic substances and reactions, and probably even of transformations. In my opinion we must recognize that we are here not only dealing with the problem of the modification of species but also with the problem of the creation of new ones." Even though science has failed to make Pasteur's hopes come true, we must be filled with admiration for the boldness of his conception of the creation of an anti-flora, of an anti-fauna, of an organically macroscopic anti-world comparable to the microscopic anti-world as it is visualized by the modern physicist. In his Utopian flights of fancy, Pasteur moves near to many contemporary biologists who, like Jean Rostand, see in biology "a positive magic" with quite immeasurable possibilities before it. So it is difficult to understand why Soviet biologists some years ago should have simply relegated Pasteur to the camp of reactionary vitalists.

2. *Stages that precede life*

In the spirit of Pasteur, contemporary science seeks to tear down the barriers which divide the different kingdoms of nature. The creation of life in the laboratory represents one of its aims, and it therefore refuses to entertain the idea that living nature can be so completely different in essence from inanimate nature that the one can be governed completely by the principle of symmetry and the other by principles of asymmetry. Like Pasteur, therefore, we are tempted to see in the crystal—as distinct from amorphous matter—the stage immediately preceding that of life, all the more so since it is here that we first encounter the *individual* in the ascending stages of nature including even right and left individuals. Nevertheless, in the view of Wilhelm Ludwig, we are not justified in trying to connect the dissymmetries of crystals (he calls them asymmetries!) with those of organisms. The former, like the latter, derive from the basic fact that in three-dimensional space there are two enantiomorphic congruent forms of every asymmetrical formation.

Moreover, there is no "continuity" between symmetries and asymmetries, even in the inanimate world. The dissymmetrical crystal is not necessarily built up from similarly dissymmetrical molecules. It was Pasteur himself who first conceived the image of the spiral staircase, which we can just as easily think of as built from symmetrical material (from regular, properly squared stones) as from dissymmetrical material (say, from irregular tetrahedrons). An example of the former is the dissymmetrical quartz crystal, whose optical activity disappears when its physical structure is destroyed either by dissolution or melting. The second is represented by tartaric acid salts; here the optical activity remains preserved even in the solution, and so betrays a dissymmetry of the molecules themselves. But that is not all. We must not only reckon with the possibility that symmetrical molecules may join together to form asymmetrical crystals (loss of symmetry); we must also bear in mind that a *straight* stairway can—with some difficulty—be constructed from irregular dissymmetrical building blocks: in some rare cases, optically active substances crystallize into regular holohedrons instead of into dissymmetrical crystals. The example of tartaric acid shows that a sum of suitable arranged local asymmetries can lead to a general symmetry (loss of asymmetry). Pasteur, it is often said, would not have failed to notice this if he had been acquainted with the work of his distinguished successor Pierre Curie—or rather if it had been possible for him to be so acquainted (cf. p. 168).

If the dissymmetry of a crystal fails to give us a true picture of molecular dissymmetry, it is even more true that the outward form of a living thing fails to reproduce the symmetry relations of its constituent parts. A snail shell which spirals to the right can quite possibly be built up from symmetrical elements, just as a spiral staircase can be built up out of square stones, or a dissymmetrical quartz crystal from symmetrical molecules. Perhaps, however, one could imagine the asymmetries of the "stage that preceeds life" making themselves noticeable in the world of the living in another way. We know that Pasteur, while working on the problem of molecular dissymmetry, discovered that it was possible to separate right and left isomers by biological means, i.e. through the activity of living organisms. Thus, by means of fermentation, he separated the two tartaric acids. The yeasts caused the "right" acid to disappear, leaving the "left" intact, because they fed upon the "right" acid. For the same reason penicilium glaucum, the fungus that has been made famous by penicillin, only attacks the dextrorotatory ammonium paratartarate. Pasteur explained this selectivity by the principle of the lock and key. "It is understandable that yeasts which have been built up naturally out of dissymmetrical substances, should find it less easy to digest nourishment which is dissymmetrical either in the same or in the opposite sense." Through the phenomenon of nourishment, therefore, the symmetry relations in the living world could be brought directly into connection with those of the chemical constitution. Indeed were we to give our imagination free rein we might well conceive that in nature, where "eating and being eaten is the supreme law", there might be whole "chains of nourishment, chains of living creatures with symmetry of a similar sense which have had their origin in one particular asymmetrical molecule. But the reality is quite different. Apart from the fact that there may be "right" and "left" substances in the same living organism, we can see from laboratory experiments undertaken in connection with Pasteur's work, that bacteria which are supplied with "right" or "left" sugar, "right" or "left" amino-acids, etc., while they seem to prefer one particular form can often also utilize for their development the form which does not agree with them. In such cases, however, the rate of growth is much slower. The right and left substances thus show no qualitative differences in their effects, though the *quantitative* differences can be very considerable.

J. Nicolle, who has been responsible for many interesting experiments on the *carbon diet of living organisms*, describes cases in which an antipode cannot serve as nourishment at all (e.g. a left(−) isoleucin). In addition, he has discovered substances one of whose forms is not only unassimilable but actually hinders growth. It is significant that he explains the disturbing effects of these inverse forms, not by the fact of their being polar opposites —that would savour of a pre-scientific way of thinking—but precisely by the close structural affinity between the normal and the inverse form. "If the most closely related substance (the optical antipode) prevents growth, then this is as though an intruder had sneaked into our house disguised as a friend, in order thus more effectively to get the better of us. One might well say that only a key that was very similar to the right key could render the lock useless, for if the key is wholly different it cannot be inserted into the lock at all."[6] We might amplify Nicolle's telling simile by pointing out that growth and the hindrance of growth cannot be thought of merely as polar opposites, for in the complicated process that leads to the building up of an organism, "inhibitions" play a very positive part. A major factor suggesting a "mythical" interpretation was of course provided by the various physiological effects of right and left substances, with which Pasteur was already acquainted. At first sight they seem undeniably to have something mysterious about them. Dextrorotatory transhexahydrophtalic acid, for instance, has a highly characteristic smell, while its levorotatory isomer has no smell at all. Geraniol gives off a scent of roses; its optical isomer smells of fresh oil. In the case of other substances we find great differences in taste. For instance, there is a strong difference between the "right" and "left" form of asparagine, which is contained in asparagus, and also between "right" and "left" sugars. A striking example of differing physiological effects of a quite different kind is provided by E. Moevus's discovery that sexual differentiation in the green alga *Chlamylomonas* is dependent on whether it is produced by the right or left form of one and the same substance. Incidentally, so far as I know, this is the only case in which the old mythical theme of a sexual polarity of right and left has been given a meaning by modern science.

The consequences of such "right-left-differentiation" can be tragic if one of the two forms is a deadly poison, while the isomer form is relatively harmless. Fortunately mistakes do not occur in Nature; those that do occur have been deliberately produced in the laboratory by the researchers eager for knowledge; thus the natural levorotatory-nicotine proves to be twice as poisonous as the synthetic dextrorotatory-nicotine. Levorotatory camphor

kills dogs and rabbits thirteen times more quickly than dextro-
rotatory-camphor. And only I-phenylalanin causes the fatal
illness called phenylketonuria. I do not believe however, that it
is possible to conclude from examples of this kind—as Hermann
Weyl obviously does—that our body contains a spirality which
is constantly turning in the same direction. It contains among
other contrasting forms the d- form of glucose and the l- form of
fructose. All these differences between right and left forms can
be explained, without assuming the existence of a polar opposition between left and right, by the principle of lock and key—
provided that we do not see in physiological properties qualities
in the scholastic sense which pertain to some "substance" or
other. There is no such thing as a sweet taste, a scent of roses or a
lethal effect. There are only interactions with the human or
animal organs of taste and smell.

3. Level of organization and symmetry

The concept of "perfection" can have many different meanings,
depending on the object to which it refers, and it often has no
meaning at all. For Plato the pentagon dodecahedron was a
very perfect body because of its qualities of symmetry. Similarly
we regard the most symmetrical of the seven crystal systems,
namely the cubic system, with its thirteen axes of symmetry and
nine planes of symmetry as particularly perfect, But the purely
formal point of view ceases to have any meaning when applied
to the amoeba. "The more perfect the creature, the more unlike
each other its parts become";[7] with these words Goethe expresses
the fact that as soon as we leave the "mineral kingdom" and
regard the form and formation of plants and animals, then the
higher we ascend the more we seem to observe a lessening and
simplification of the symmetries. Goethe's view is upon the
whole correct, but we must not imagine that living creatures

form a progressive chain whose last and most perfect link is provided by the vertebrates with their relatively low right-left symmetry.

Amongst the lowest forms of living organisms there are many highly symmetrical ones which furnish a considerable portion of the *pattern in Nature's art*. (Anyone who has no opportunity to look at Ernest Haeckel's famous work or D'Arcy W. Thompson's *Growth and Form* will find illustrations in the work of Herman Weyl.) It is true that spherical symmetry is never completely attained as in the case of crystals. But in the animal and vegetable kingdoms we frequently find that "symmetry of five" which is wholly lacking among crystals—as witness the pansy and the starfish. Symmetries which are peculiar either to animals or to plants are completely unknown. Above all, we cannot say that the imperfect symmetries, namely the radial, bilateral and dorsiventral ones—actually those of the human body—only come into being at a certain level of organization and then replace the more perfect symmetries of the lower orders of life.

According to L. Wolff and D. Kuhn, as perfection increases symmetry is progressively limited by the "biological phenomenon of polarity". But most biologists reject this idea, which, after all, comes very near to the idea of a positive tendency towards asymmetry. There is no need to introduce a special factor which works against symmetry; *asymmetry is not the opposite but the absence of symmetry*. Phylogenetic considerations can explain in each individual case why symmetry vanishes as *function* gradually takes precedence over *form*.

Let us look at Plato's myth in the Symposium (cf. p. 15), with its many interpretations, as a symbol of *phylogenesis*, that is to say a symbol of the development of the species as distinct from the development of the individual. Unquestionably, our androgynous spherical-symmetrical ancestors must have had difficulty in moving in a straight line—according to Plato they were compelled to turn cartwheels—and it was thus very much to their advantage that, thanks to Jupiter's wrath, they had to adopt the less perfect but, from this point of view, much more practical right-left symmetry. *The connection between bilaterality and locomotion is* really almost self evident. To express the matter in finalist terms, the living organisms became right-left symmetrical, or dorsiventral, in order to be able to move about better. We can see this in animals like the sea-urchin which, as stationary adults, are endowed with spherical symmetry but have right-left symmetry in the more progressive and mobile stage of larvae. A particularly good illustration is given by Wilhelm Ludwig: the starfish is descended from bilateral ancestors which adopted a fixed position and in due course acquired the radial body form (symmetry of five); "but this radial body form is only an outward characteristic, and when some starfishes recently reverted to a mobile form of life, a second bilateral symmetry came into being which ultimately led to elongated, worm-like forms".[8]

But movement is not the sole function of an animal and on a still higher level even the right-left symmetry which is a "lesser" symmetry in relation to other symmetries can prove obstructive. The highest forms of organisation cannot be realized at all through the aesthetically satisfying form of strictly regular construction. As A. Portmann shows in his book *Die Tiergestalt* (the Animal Form), among the vertebrates which still have an outward appearance of "bilaterality", it is only the empty body, from which all the intestines have been removed, that can really be divided into two enantiomorphic halves. "It is true that here too the internal organs almost always come into being symmetrically, but this principle of construction is discarded at an early stage in favour of another which ensures the complete use of the restricted space inside the body cavity. The internal organization becomes unsymmetrical."[9] In these circumstances it is not surprising that, in dealing with a question which is raised in this connection in the *Studium Generale*, the zoologist (Wilhelm Ludwig), unlike the botanist (Wilhelm Troll) should have stated that considerations of symmetry hardly play any part in the realm of living things. Static conditions, interpreted in their broadest sense as conditions of selection, are for the most part supplanted by differential equations, in which time is a variable.

These provide "flowing equilibria" in which the character of the past and of the historical finds expression. Ludwig therefore takes the view that for his science, zoology, there has not as yet been any need for superordinate principles. Selection, environment, chance, determine in individual cases the relations of symmetry.

4. The significance of right and left in the animal kingdom

The physicist is not greatly interested in the spherical symmetry of the earth, which seems to him to be "natural"; it is precisely the deviations which give him valuable knowledge (for instance the flattening of the poles tells him that the earth turns around its own axis). The same holds good in biology—if we disregard "idealist morphology". Hence, as Hermann Weyl points out, Wilhelm Ludwig pays hardly any attention, in his monograph on the right-left problem, to the origin of bilateral symmetry, and goes into great detail about the secondary asymmetries which are superimposed on the symmetrical ground plan. Why, he asks, should there be asymmetry at all? Why right rather than left? (Or left rather than right?) He asks these two questions because of the numerous deviations from right-left symmetry which can be observed in bilateral animals. Anyone wishing to form an opinion on the significance of right and left in the world of living creatures must read this thorough yet wide-ranging book. Here, of course, it is impossible to reproduce even a small selection of the examples which Ludwig quotes; they range from all manner of asymmetrical characteristics, both morphological and physiological, to the asymmetries of habit. The former include pairhood and non-pairhood of organs, the direction of the convolution of snail shells, heterochelia of crustaceans, i.e. the greater development of one claw in the lobster and in the male fiddler crab, where this claw is a secondary sex characteristic used for fighting or for waving.

They also include the asymmetrical pairing organs of the *Zygonectes*, the crossing of the beak of the crossbill, the formation of antlers, and asymmetrical markings and colourings in general. It is precisely these formal asymmetries in the animal kingdom which are so connected with those wonderful teleological adaptations and co-adaptations which provide the vitalists and neo-vitalists with their arguments. Thus hermit crabs are adapted to living in shells with a right-hand convolution. The wings of the longhorn grasshopper form "musical instruments" because the right wing has a stridulating organ which by being drawn across the ribbed scraper of the left wing emits a sound.

The asymmetrical habits of many animals approximate particularly closely to the bilateral human asymmetry, which is mainly functional. For instance, there are the pairing habits of spiders and zygonectes, combined in the latter case with morphological asymmetry; the leg-lifting of the male canine; the "loping" of wolves; the gallop of the horse (by nature usually a left-hand one but becoming right-hand after being ridden in—by a right-legged rider?); the leftward habit of attack among stags and wild oxen; and the trajectories of a number of animals. The "lying habits" of flat fish have formed the subject of many searching enquiries but it is still an open question whether the plaice has a crooked mouth because it always lies on one side or whether it took on this habit because of its asymmetrical shape. We really only speak of handedness in reference to rats and monkeys and it should be noted that the latter are accounted as ambidextrous (cf. p. 111).

Some of these right-left asymmetries are *racemic*, which means that the relation between left and right is approximately 50-50. Thus about half the crossbills cross their beaks from right to left. Others are *monostrophic*, which means that more than 90 per cent are on one side; thus among a thousand specimens of a particular species of snail (*Helix pomatia*) only one will be inverted. In a third category the distribution lies somewhere between these two extremes, and it is of course this *amphidromous* category that is by far the most interesting for the man of science, since the relation between right and left cannot, as in the case of racemia, be explained by the operation of chance and—apart from this—the proportion of the "inverse" is too great for it simply to be treated as a "misprint" in the process of heredity.

From all the wonderful facts and all the observations which he had made and collated in his book Ludwig inferred—and for the lover of philosophic speculation the inference must have been disappointing—that we know of no law governing any general distribution between left and right in the animal kingdom. Indeed "the determinant for the distribution of right and left in the animal kingdom is what we generally call chance".[10] According to this view there is no such thing as a uniform right-left orientation, no preference for one side over the other.

The importance of Ludwig's statement cannot be exaggerated. By posing the problem of right and left as a problem of *distribution*, he causes it to disappear. If the tails of all dogs curled to the right and the shells of all snails had a right-hand convolution, if all plaice lay upon their right side, all insects fiddled with their right wing, if hops twined round the pole from the right like all other climbing plants, then this—together with the right-handedness of men— could cause us to seek a reason for Nature's preference for one particular side. But since things are not like that, there is no point in any attempt to deduce a general rightward orientation—say from the rightward twist of our solar system or from the chemical constitution of the earth. Nor can we adduce the distribution of right and left in the animal kingdom or that of organic right and left substances to support a theory of the Pascual Jordan kind concerning the origin of life —i.e. conclude from a supposedly non-fortuitous distribution of right and left that the beginnings of life must have been due to a singular and improbable occurrence. If such theories are correct, they will have to be vindicated by some other means. It should be noted that not only is there no proof of any general orientation on the part of nature, but that it *cannot* be proved. Ludwig has shown how the problem of establishing homologies is posed at the level of asymmetries of form. He distinguishes four great "classes of asymmetry": bilateral bodies with unequally developed secondary characteristics; screw-shaped and turbo-spiral forms, (snail shells); multiloculine (*foraminifera*) and circular spiral or cycloid forms which can only be seen from above because they are two-dimensional (trajectories of moving animals). *Every time we pass from one class to another, right and left must be defined anew.* For instance, it is in no way established through comparison with our own body pattern that a clockwise movement should be described as orientated to the right. Even within the same asymmetry class an unlimited homologiation is hardly possible, according to Ludwig. If we had the good fortune to dig up a hand of the Venus de Milo we could tell whether it was a left or a right hand. Nor would there be any doubt about our dog's paws, and if we take the direction of movement into account we could transfer to certain swirling infusoria the nomenclature derived from the morphology of the human body. But here we reach the limit of the possible. When we come to such matters as the *situs inversus* of the trematodes or the position of insects' wings, new definitions are required.

5. *Right-left asymmetry as a mark of species*

Once pseudo-philosophical speculation has been eliminated, all scientific problems do not of course disappear. We are still left with the problem of the constancy of asymmetrical characteristics within the species: the quite remarkable fact that the relation between right and left is a characteristic of an animal species (so characteristic in fact that in the case of well known monostrophic groups like the *foraminifera*, specimens that are inverse but otherwise indistinguishable from the others are regarded as members of a different species). This problem is obviously a part or aspect of the general problem of the constancy of the characteristics of species. Apart from this does it present any special difficulties? Suppose that the problem of the auto-reproduction of the genetic material had been solved in principle and that it was no longer at all mysterious. Would there then still be a right-left problem in addition, or are asymmetrical characteristics transmitted in the same way as symmetrical ones?

Let us take as an example of a comparatively simple heredity mechanism the "mass production" of little medusae by budding, D'Arcy Thompson describes it. The parent animal so he tells us, is a vortex-like bell with a symmetrical handle. "No sooner made, then it begins to pulsate; the bell begins to 'ring'. Buds, miniature replicas of the parent organism, are very apt to appear on the tentacles, on the membrium, or sometimes on the edge of of the bell. We seem to see one vortex producing others before our eyes. . . . Certain it is that the tiny medusoids of Obelia are budded off with a rapidity and a complete perfection which suggests an automatic and all but instantaneous act of conformation rather than a gradual process of growth."

If the manufacture of young snails, fiddler crabs, or even of human beings, proceeded as quickly as this, we should perhaps have no more difficulty in imagining the transference of asymmetrical characteristics, than of symmetrical ones. But the process of inheritance has not only this steroyped and automatic aspect which we see in the case of *budding*. In *sexual inheritance* the parent organism generally produces an egg, a germ cell that is in no way similar to it. The process of *growth*, which in the case of the medusa has been as it were concealed by the speed of the operation, contains the whole problem of heredity—the relation of genotype and phenotype, the nature of "biological traits". The question arises as to what is actually inherited. A substance? An ordered substance? A system? Here there arises a special problem of bilateral asymmetry which complicates the great riddle of heredity and at the same time illuminates it in a rather curious way. Are asymmetrical characteristics inherited as such? Or are symmetrical characteristics inherited asymmetrically? (cf. p. 126).

The ontogenetic process which to us appears most mysterious of all is that of *cell differentiation*, the fact that although the division of the egg-cell in regard to the chromosomes is *equal*, the daughter cells of the newly-formed organism inherit different properties and capacities and show signs of a different pattern of organs. The direct and the indirect methods of the research scientist here seem to give contradictory results. The microscope shows us an evenly formed egg plasma, in which the two nuclei are symmetrically arranged. But the refined experiments of the developmental biologist point to a hidden "prospective potency" in what is apparently something quite unformed. One part of the cell was destined to become mesenchyme, the other, indistinguishable from the first, was not so destined. Hans Driesch saw the reason for this in an inequality that was already present, the anisotropy of the plasma, but he held that we only knew of this anisotropy through its subsequent appearance. "We could not foretell from the fact that we had before us a blastula with similar nuclei and an anisotropic plasma construction, that this blastula at one pole of its axis would demonstrate processes of cell partition and growth."[11] He regarded the *anisotropy as a cause that was not absorbed into its effects*. For him the localized potentialities are "poles" in the mythical sense.

Modern research is concerned to make of the anisotropy of the plasma a "normal" cause which is resolved into its effects. It transfers the study of cell development into the realm of chemistry. By means of greatly perfected instruments (the electron microscope) it examines the fine structure of proteins, chemical differences of the plasma, catalysts, differences in pH, in short everything which makes the apparently uniform egg a complete laboratory with very complex equipment. The anisotropy of the egg thus becomes a reality. Incidentally we, now know that in the case of certain animals, such as the seasquirts the anisotropy is visible to the naked eye as an asymmetrical arrangement of parts. In others it can be made visible by means of special marking techniques (Vogt, 1925). A ring of red pigment will then appear from beneath the equator towards one of the poles of the egg (the vegetative pole) which itself remains a bright apical area free of pigment.

The anisotropy of the egg explains in principle the differentiation, that is the differences in behaviour between the two poles, of which the one, (the animal pole) subsequently produces mesenchyme while endomesoderm tissues are formed from the other. The explanation, I say, is in principle, for the details of the differentiating process are at the moment still largely unknown. But certain interesting experiments show the biologist that he is on the right road and that the differentiation within the cytoplasm of the egg is certainly of a chemical and quantitative character. The entire morphology of the sea-urchin can be completely reversed by a relatively simple chemical treatment of the germ-cell. Lithium salts produce vegetativization, i.e. the predominance of the endomesoderm tissues, while conversely sodium cyanide leads to animalization.

In three different ways, during the progressive differentiation of the growing organism, the locus becomes evident: as *polarization* = the construction of axes at a very early stage (promorphosis), as a *localization* of later organs and as *lateralization*. The chemical anisotropy and the interlocking of the various mutually adjusted processes in which each phase creates the necessary conditions for the next, are all well designed to explain to us the localization which is as it were imparted with "the localized". Less clear but still apparently explicable is the formation of axes during the promorphosis, the alignment of the material particles, possibly under the influence of the maternal organism (Goldschmidt). The concept of lateralization remains the most mysterious of all. If a biologist speaks of a lateralized or an "already lateralized" germ cell (cf. p. 125) he does not merely mean that the germ cell possesses a right side or a left; he is also thinking of the potential determination of a pattern. The future heterochelic lobster does not have a large claw on the right but a right large claw, as though the "right" were a part of the information provided by heredity.

If lateralization is regarded as nothing more than a localization, this is due to the curious results of experiments in the regeneration of organs and entire complexes of organs. Ludwig first cites a celebrated experiment which we owe to Hans Spemann. A triton embryo is divided lengthways at the beginning of gastrulation or earlier (Spemann carried out this delicate operation with the hair of a child). The halves of the embryo then complete themselves to form a somewhat smaller but *complete* individual, and we notice the peculiar fact that in the animals produced by the left halves the internal organs are arranged in a normal fashion while 50 per cent of those from the right half display *situs inversus*. Ludwig's second example concerns the "phenotypic inversion" which can be produced in the young lobster by amputating the right claw, which is normally the larger one. A new one then appears, but the relations of size are changed and the creature becomes "left-clawed". Ludwig infers from these two experiments that the tissues must be regarded as the carriers of asymmetrical properties; not that the actual potentiality of a "right large claw" is included in them— such a "substantialization" would indeed be shocking—but they do "contain" a right agency and a left agency as it were. These are distributed throughout the organism with a certain concentration gradient in such a manner that the concentration of the one inclines from right to left and of the other from left to right, and they direct the deployment of an asymmetrical formation towards a right or a left form. As a non-biologist Hermann Weyl prefers to withhold judgment on these matters and I follow his example for the same reason. I would merely remark that the idea of such a concentration gradient tempts one to think in terms of "quantities" of right and left, as though something could be more or less right or left. It must be admitted that this is a disturbing thought. I do not know whether this theory of agencies has been subsequently sustained. We can only hope that a simpler explanation can be found for the phenomenon of claw reversal in the lobster. As to Spemann's triton embryos, the well-known medical term of *situs inversus* may here prove tricky: to have one's heart on the right side is not the opposite of having it on the left, but of not having one's heart on the right.

V

THE LATERALITY OF MAN

1. What is a right-handed or left-handed person?

When we leave the plant and animal world, with its inexhaustible variety of form, we must obviously change our perspective completely. Form yields place to *function*, or rather, if we follow modern trends in thought and reject the idea of a crude contrast between form and function—the formal aspect yields place to the functional. The human body possesses no refined highly specialized "one-purpose tools", but it has a single all-purposes tool—the hand. We have a right and a left hand which scarcely differ from one another in form and size, and yet differ exceedingly in regard to function. It is obviously this disproportion between the symmetry of form and the asymmetry of function which turns the handedness of man, and, more generally, his laterality, into a genuine scientific problem. If—as might perhaps be the case with the inhabitants of Mars—we had a pincer-shaped limb on one side and a hammer-shaped limb on the other, we should treat it as a matter of course that we put each of these limbs to a different use. And yet, when we think about it, we are startled by this difference between two symmetrical organs which are to a very great extent similar in form and which are obviously designed for the same uses. Why is one of these members more suited to certain actions than the other? There is nothing about our right hand to suggest that it writes better than the left. Every definition of handedness must take these facts into account, it must therefore be an *operational* definition.

It has been said that there are as many definitions of handedness as there are theories, and actually it is almost impossible to define what exactly constitutes a left-handed or a right-handed person without anticipating an explanation of the origin and nature of handedness. If we speak of the "preferential use" of one hand, this suggests a conscious choice, a cultural or social origin of handedness. The expression "natural physiological aptitude" leads a *priori* to acceptance of the opposite view. If, as many authors claim, we are referring not to the hand but to the dominant half of the brain (cf. p. 112) in giving this definition, we are of course presupposing the existence of a dominant half of the brain. Further, which actions should constitute the criteria for handedness? All? Even those where, through simple force of habit, only one side of the body is involved, as in folding one's hands and shaking hands, actions which might just as well be undertaken by the other hand? Perhaps we should also include involuntary movements and the "elementary movements". It is obvious that if, like the majority of modern scientists, we insist on confining the concept of handedness to certain definite highly differentiated skills with a specific purpose behind them, this does away with certain oversimplified theories concerning the origin of the phenomenon. On the other hand, writing, playing the fiddle, driving in a screw—to name but a few of the examples quoted in the literature on the subject—are accomplishments that neither a child nor an animal can acquire. By thus narrowing our choice we tend to erect an impenetrable wall between man and the animals, and thus perhaps make the understanding of the ontogenesis of handedness more difficult than it would otherwise be. The scientist, however, will naturally wish to give this phenomenon its place in the development of the individual as well as in the evolution of the race.

The question whether laterality is an exclusive prerogative of man is answered in various ways. Although asymmetrical "habits" are widespread in the animal kingdom and many of them are racemic and constant in the species, it would be an exaggeration to see in this more than a vague distant anticipation of the far-reaching specialization of the two halves of the human body. These habits are often morphologically conditioned or go hand-in-hand with a morphological asymmetry, and where this is not evident we must suspect its presence. Thus when Ludwig tells us that crickets—with the exception of the ambidextrous mole cricket—cannot chirp once the right-hand wing happens to get "derailed" under the left (and this despite the fact that their stridulating apparatus is well developed) then, surely, one tends to think of some hidden organic cause rather than of an insufficient skill on the part of the fiddler. It is also surprising that our nearest relative in the animal kindgom, as Wolfgang Kohler's experiments on monkeys show, is ambidextrous.

Probably the functional asymmetry of man is directly connected with the process of "hominization", whether it is hand or brain that plays the principal part in this matter. It is thus the last achievement of phylogenesis—so far at least; for I should not omit all mention of the curious idea of the biologist Jean Rostand that an artificial strengthening of laterality— possibly through the multiplication of brain cells on one side— may produce an improvement of human faculties and turn us into "supermen".

2. *Laterality—a central phenomenon*

If the phylogenesis of right-left asymmetry leads us to the "white spot" of hominization, ontogenesis permits us to enter the safer territory of experimental research. Yet we are still removed from a satisfactory theory about the nature and origin of laterality in the individual. In a certain sense such a theory would actually provide a solution for the most profound and complicated questions that man can pose. There are, however, a few matters on which the majority of present-day research scientists are able to agree, after centuries of error largely induced by prejudices concerning right and left.

The most important fact, concerning which hardly any doubts are entertained nowadays is that *laterality constitutes a central phenomenon*, that is to say a phenomenon whose seat must not be sought in the peripheral organs but in the *brain*. Broca's remark[1] that we are right-handed because we are "left-brained", is confirmed today by numerous clinical investigations which have shown that functional disturbances are connected with the opposite half of the brain. The better functioning of the right hand depends on the better functioning of the left half of the brain. Hence, one might add, it is precisely in the "nobler" organ—contrary to the popular belief that right is better than left—that the left side is the "more excellent".

Following the French neurologist's important discovery, the idea began to gain ground—once various obstacles had been overcome—that man is not ambidextrous by nature but that the asymmetry (of the brain, or of the brain functions) is innate in him. Another new concept was that of a general brain-directed laterality, affecting not only the hand but also the nether extremities and even those sense organs which are arranged in pairs. In the case of the last-named, however, it became necessary to frame new concepts. Following the analogy of handedness where skill, and not mere physical strength, is the determining factor, it was decided to regard as "left-eyed" not the man whose sharpness of vision is stronger on the left, but the man in whom, of the two images that are projected on his retina in binocular vision, the right is suppressed in favour of the left. It has not been possible to give conclusive proof of complete laterality in human beings, but most scientists today take the view that a normal man, if he is a definite "rightist", is oriented to the right from head to foot and to the left from head to foot if he is a "leftist". Where the facts do not confirm this, it will be suspected that the person in question does not display his "proper" handedness but has changed it to compensate for some damage to the central nervous system, which may often have been caused in intra-uterine life. So the conception of handedness as something brought about by the domination of one half of the brain has given rise to a distinction, which is universally recognized today, between genuine and pathological right- or left-handedness, quite apart from that caused by re-education.

This distinction which, from the practical point of view, was certainly the most far-reaching result of Broca's discovery, failed to achieve immediate acceptance and ultimately only did so with difficulty. Even at the beginning of this century, the combined effect of the prevailing belief in the omnipotence of heredity and the old prejudice against the left, resulted in the dividing line between normal and abnormal being drawn quite differently from the manner in which it is drawn today. Only the right-handed man was normal, while left-handedness was regarded as a morbid, degenerate symptom and was connected with every imaginable kind of defect, not only of a physical but of a pyschological, even of a moral kind (cf. p. 133). The name that is principally associated with the alleged discovery of a pyschophysical syndrome among the left-handed is that of Lombrose.

Here we have an example of the crippling and damaging effect of the ancient prejudice against the left even upon exact experimental science. This false, "Manichaean" differentiation disguises the real difference between normal and pathological laterality, and obstructs the search for the really effective causes. It is quite true that more left-handed people are to be found among stammerers, epileptics and the mentally afflicted in general than among the healthy section of the population. But this fact, which was emphasized by Lombroso, is explained by the fact that the same cause, such as an injury to the brain for instance, but also mere re-education (cf. p. 135) in some cases, can have various consequences such as stammering and epileptic fits, and at the same time produce a reversion to the use of that half of the body which was originally not the dominant one. Conversely, however, we do not find more stammerers, epileptics and so on among normal left-handed people, in whom no such additional pathogenic cause has been operating, than we do among naturally right-handed folk. . . . The error in reasoning lies in mistaking an effect (left-handedness) for a cause. It is obviously favoured by the unequal distribution of laterality. The result of this is that the secondary organic or social causes produce many more pseudo-left-handed than pseudo-right-handed people.

The new discoveries had a direct effect on medical and educational practice. If there are naturally left-handed people, if left-handedness is not in itself a defect, if left-handed people are not "inferior", and if nevertheless artificially right-handed people are really being forced to some extent to act contrary to their real nature, the question naturally arises as to whether the process of so called "*re-education*" of left-handed people, which has always been practised to some extent, does not constitute a serious and unjustifiable form of interference.

In this connection *the relation between laterality and speech* plays a very important part—above all the discovery of a speech centre, the crown of a whole series of discoveries of important "localizations" in the brain. It is held that the speech centre, unlike the motor and sensory centres, is *unilaterally* localized and, in the case of the right-handed person, is in the left half of the brain. Broca said quite positively that we "speak with the left half of our brain". Today people are less categorical about this, yet it is generally recognized that both halves of the brain are not of equal importance in regard to the functions of speech. The fact that unilateral damage to the brain can rob a man of speech proves this beyond a shadow of doubt. It is thus only to be expected that among people who have been robbed of their natural laterality—among those who are pseudo-left-handed or right-handed owing to pathological causes and also among left-handed people who have been re-educated to right-handedness —there should be a disharmony of neurological functions and particularly an increase of speech defects. It sometimes happens that with such people "the two halves of the brain quarrel as to which of them is to be in control". This argument, which is brought forward by Froeschel, Orton and Travis, Norbert Wiener and others against re-education, is confirmed by a great quantity of statistics; it has also been proved that stammering generally occurs at that exact point of time where the child, if it is not right-handed already, is forced by its surroundings to become so.

Thanks to the discovery of cerebral dominance, research into brain activities has begun to move along new paths, and our knowledge of this field has become much more profound. Yet it would be a mistake to believe that the true nature of laterality, i.e. of the functional asymmetry of the brain, is now understood; only a few scientists believe that the slight asymmetry of this organ, as it manifests itself externally, can really provide us with an explanation of laterality. There is nothing in either hemisphere—size, structure, volume, number of brain cells, complexity of the convolutions, efficiency of blood supply, excitability, and so on, that has not by now been measured with the most delicate methods known to modern science. Yet it has proved just as impossible to discover in the brain a material basis for the functional differentiation of its two sides as it was in the case of the peripheral organs. We are told that the right brain of Ludwig Edinger, the left-handed American psychologist, was larger than the left and was particularly well developed. In general however the differences are unimportant. Even the results of electro-encephalographic examinations have proved to be irrelevant. We are thus concerned with localizations without knowing what has been localized.

No scientist today would dispute the view that gestures, which we usually observe under their visible peripheral aspect of muscle, are primarily inner-cerebral central processes. And the basic idea of the difference between the two halves of the brain still continues to be regarded as valid. But the beautiful simplicity of Broca's analysis has long since vanished. A person who has lost the power of speech owing to brain injury very often regains that power after quite a short time, and the self-regulation of the nerve apparatus takes place even more quickly in cases of *apraxia* than in cases of *aphasia*. It is well known that those who have been crippled or amputated on the right side learn very quickly to use the left hand for writing; in fact after overcoming the first inhibitions, they discover that they have always been capable of using it. At all events, they do not have to learn to form letters with the left hand as they did in childhood with the right. Apart from Pasteur (cf. p. 91) we have the remarkable example of Auguste Forel who lived on for nineteen years after a stroke and during that time wrote his five-volume work on ants, amongst other things, with his left hand. We infer from this that the weaker half of the brain can, if necessary, take over the tasks which were originally allotted to the stronger half. Each cerebral hemisphere not only controls the opposite half of the body but also exerts, though to a lesser degree, an influence on its own side. According to Paul Girard the rarity of aphasia in cases of right-sided cerebral paralysis of the brain among children proved that the dominance of one hemisphere of the brain is not exclusive. *The brain works as a whole.*

Twenty years ago the knowledge that there was no hope of finding a material basis for laterality would have confronted science with an insoluble problem. Today *cybernetics* provides us with the possibility of steering a course between the Scylla of mechanism and the Charybdis of "spiritualist theories". The "model" of the brain is not the machine in the primitive mechanical sense but the self-regulating thinking machine, the computer, which operates on the basis not of an empty physical structure but of the co-operation of such a structure with the instructions fed into it initially and all the other instructions received from the outside in the course of a chain of operations and stored in its "memory".

Recent works on cerebral dominance draw their conclusions

from the facts thus revealed. According to O. L. Zangwill, the human race can no longer be regarded "as divided into two mutually exclusive categories—sinister and 'right-brained', dexter and 'left-brained'. Handedness must be regarded as a graded characteristic, left-handedness in particular being less clear-cut than right-handedness and less regularly associated with the dominance of either hemisphere. Indeed, cerebral dominance is in all probability itself a graded characteristic, varying in scope and completeness from individual to individual."[2]

The views expressed here, which have been deduced from clinical examinations, are complementary to the results achieved along quite other lines by Arnold Gesell. Using such aids as *Cinemanalysis* and other modern techniques, the American psychologist in the Yale Clinic of Child Developement has studied the manifestations of laterality at various stages of development. The "ontogenesis of child behaviour" leads, if all right-left education is excluded, not to ambidexterity but to functional asymmetry, since, according to Gesell, "effective attentional adjustments require an asymmetric focalization of motor set". *All normal children grow up to be right-handed or left-handed.* In the course of this process the child passes through a number of stages in which there is an alternation not only between right-handedness and left-handedness but between symmetry and asymmetry. "The bi-polarity which bisects the Emersonian universe also bisects the organization of laterality in a growing organism. Two pairs of opposing trends are in mutual rivalry: bilaterality versus unilaterality, and right versus left. This gives rise to many inflections and combinations. These trends are probably not in actual conflict or in true competition. They are in developmental flux."[3]

Gesell emphasizes that the problem of handedness has been greatly over-simplified and that we are here concerned with a very complex trait, which is intimately bound up with the total action system of the child. It is, as he says, a product of growth. One can perhaps blame Gesell because by this definition he evades the necessity of making up his mind whether handedness is inborn or acquired, and in fact Gertrude Hildreth states that in his later writings he left this question open. Yet in dealing with a trait which only "becomes a reality" during the course of life—a new-born child is ambidextrous in the same way that new-born Dalmatian puppies are white—it is impossible to avoid the concept of growth. It is certainly not an easy one to handle, containing within itself, as it does, all the unresolved problems of structure and behaviour. Yet modern science needed it as a contrast to the concept of development. While the latter refers to the constitutive elements which are characteristic of the species, growth and maturing bring about the progressive differentiation of experience and its integration into the individual. It should also be pointed out that by regarding right and left laterality as a product of growth we are for the first time applying the perspective of history and introducing the factor of time. Functional asymmetry thus acquires the character of something that has "come into being". But does that mean that we can no longer regard it as something *congenital*? Certainly not. Gesell seems to take the view that decisive phases of the process of growth occur in the pre-natal stage. He devotes considerable attention to the *tonic-neck* reflex[4] discovered by Magnus, the earliest known expression of laterality in a little child. This reflex can be observed in a twenty-eight week old embryo. It shows a similar "amphidromous" distribution to that of laterality, and Gesell suggests that we should in fact see in the orientation of this remarkable phenomenon an indication of the later decision: "right or left."

3. *Two famous theories concerning the origin of laterality*

Most of the older theories concerning the origin of laterality appear to have been finally discarded in favour of the theory of central dominance—the purely social theories, for instance, and also the idea that the position of the embryo in the uterus is the sole determinant of the nature of the ultimate laterality. Nevertheless I propose to examine two of these theories because they have been particularly dear to myth lovers and have played a certain part in the history of thought.

Plato's explanation of right-handedness—it is significant that he does not mention left-handedness—is governed by the *myth of ambidexterity* (cf. p. 145). He fails fully to grasp the idea that "right and left, so far as the hands are concerned, should according to nature be different, in all activities, whereas in the case of the feet and the lower limbs in general, there is no such difference but both are equally strong whatever the effort demanded of them". Nevertheless, he continues, "In our hands we have all—through lack of understanding on the part of our mothers and nurses—been made equally lame, and whereas by nature these two limbs are more or less of the same strength, we ourselves have made them different in so far as by reason of prevailing customs we have not used them correctly" (Laws, Book 7). According to Plato, the lack of understanding on the part of mothers and nurses shows itself not, as it does today, in attempts at re-education, but in the practice of carrying the children on the right arm, as a result of which the left arm of the child is pressed against the body and is thus hindered in its development. Apart from the fact that Plato is here in error—children are carried on the left arm so that the right hand may be left free—there is a fundamental flaw in his theory; it takes no account of the numerical relation between right-handed and left-handed people. Indeed it does not even answer the question why left-handed people exist at all.

According to the Pye-Smith-Weber *weapon theory*, the precedence of the right hand can be traced back to an early epoch in the story of mankind and is connected with the use of weapons. This is an idea which Carlyle also expresses in his diary (1871): the heart, which lies on the left side of the body, was turned away from the enemy and protected by the shield that was held over it, while the right hand wielded the weapons; thus the right hand because the more skilled, undertook the most difficult part in all actions and so induced right-handedness. This theory, if we overlook the questionable character of its etymological arguments (cf. p. 33), can claim, support from the fact that in antiquity handedness is nearly always mentioned in reference to the use of weapons (cf. p. 134). But it is only meaningful in connection with a theory of heredity which explains how this habit had been passed on to us and why everyone is not right-handed. For the very rare occurrence of the *situs inversus* cannot explain the existence, that is to say the survival, of left-handed people. One might say that any such line of reasoning assumes the validity of the theory of the inheritance of acquired character-istics, which, in general, science rejects. But at the turn of the century this argument was not nearly so decisive as it is today. And one can understand how it was that the weapon hypotheses continued to give satisfactory answers for a very long time to all the questions connected with right- and left-handedness. It can even be harmonized with the theory of cerebral dominance. The right-handedness induced by the use of weapons induces in its turn left-brainedness!

The unsatisfactory character of both the two hypotheses sketched above is obvious enough. If it is to be of any use at all, a theory of laterality must begin by explaining the distribution of left-handedness among the human race. Why are we not all ambidextrous, i.e. why do left-handed people exist at all in any numbers worth mentioning. Why is the distribution not racemic, that is to say, why are there not approximately as many left-handed people as there are right-handed people? For unlike the "mythical" question "why precisely right and not left?", these are reasonable questions.

As to the number of left-handed people in proportion to the total population of the earth, estimates differ widely; they vary between 1 per cent—only in the case of older writers—and 30 per cent, depending on the criteria used. The assessment of the percentage of left-handed people results from questionnaires and more recently from tests—the Merril-Palmer test, the Harris test, etc.—which may or may not involve skills that had to be specially acquired. It is also said that we have been able to trace a larger number of left-handed people since being left-handed ceased to be a "disgrace." The exact figures are not of great importance. The actual number of left-handed people is certainly not so small that left-handedness should be regarded as an "anomaly", while on the other hand the distribution is not racemic. Chance is therefore ruled out in two of its forms: as a mistake on the part of nature and as the working of blind fate.

4. Is left-handedness hereditary?

It is usual to divide theories of laterality into three large groups, according to whether we regard heredity, or the external environment, or even pre-natal conditions—such as the position of the embryo in the mother's womb—as the decisive factor. In former times, environmental theories carried the most weight, since the concept of heredity was not yet fully developed. In our own day there are only a few scientists—e.g. Gertrude Hildreth[5] —who completely exclude heredity. But the contrast between heredity and environment has lost most of its sharpness. The defenders of the heredity theory point to the difficulty of explaining the relatively high percentage of left-handed people in human society in any other way. They can also point to the early appearance of laterality (according to some scientists a pre-natal appearance—in the "tonic-neck reflex") and to its persistence, which is said to be in evidence even in the last twitches of the death agony. According to G. H. Hewes, the genetic explanation of handedness is now on a fairly solid foundation, thanks to recent research which has established a significant correlation between functional asymmetry and the morphological symmetry appearing in the dermoglyphes, especially in the patterns of finger-prints and hand-prints.

The decisive factor—in the absence of experiments, which cannot be carried out in this sphere—is of course research into family statistics. According to David C. Rife's classic work, pronounced left-handedness appears in only 7 per cent of the children of right-handed parents, in 20 per cent if one of the parents is left-handed and in 50 per cent if both of the parents are left-handed. Numerous other works of this kind make it seem highly probable, at least, that factors of heredity play an important part in establishing laterality and especially handedness. No one, however, accepts these as simple mendelian traits. The "laterality formula" as such is not transferred to the children from the father or the mother as happens in the case of blue eyes; it requires the joint operation of numerous genes. Wilhelm Ludwig conceives of a mechanism for the inheritance of laterality somewhat on the lines of Richard Goldschmidt's well-known theory of the inheritance of sex,[6] according to which the control of the inheritance-factor-equilibrium is actuated by the X chromosome mechanism. It would actually be possible, he declares, to replace the "male" and "female" in Goldschmidt's formulation by "right" and "left" or "left" and "right".[7] The advantage of Ludwig's theory lies in the fact that it accounts for the graded nature of laterality and the existence of ambidextrous persons. But, understandably enough, it lacks anything analogous to Goldschmidt's beautiful experimental confirmation through the artifical displacement of the equilibrium and the generation of "intersexes". Above all, it necessitates the introduction of a mysterious agent—something which is always liable to arouse our suspicions in science—and points to an insufficiently precise establishment of the facts (cf. p. 108).

Although the hereditary nature of handedness is now regarded as proved, we know hardly anything at all about its mechanism; something is inherited, but we do not know what that something is. Our ignorance is twofold. The inheritance of an asymmetrical function poses, on the one hand the problem of the inheritance of *asymmetrical* features, and on the other the problem of the inheritance of *functions*. The science of genetics has still not

solved either of these problems—and particularly the second—
because, successful though it is where physical properties are
concerned, its conceptual apparatus, which is highly suitable for
dealing with the laws governing the transmission of material
genetic entities, appears to be sadly inadequate for the purposes
here in question.

It is particularly "annoying" for anyone convinced of the
hereditary character of laterality that the *twin method*, that
favourite tool of the student of heredity, appears completely
useless in the present instance. One-egg twins, which, to some
extent by their very definition, are held to possess the same
hereditary equipment, may well not have a similar laterality;
and one may be right-handed and the other left-handed.
Moreover other asymmetrical characteristics, both morphological and functional, can appear in reverse directions or, as we
say, in mirror-image. Zazzo describes two little twin sisters who
come into this category. (One-egg twins are always of the same
sex!) "Not only was one of them right-handed and the other
left-handed but, when I examined their motor behaviour, the
gestures made by the one were mostly reversals of the gestures
made by the other, and—here comes the culminating point of
this impression of a mirror-image—both the twin sisters showed
the same convergent strabism, in one case of the left eye and in
the other of the right.

5. *The paradox of mirror-image twins*

We must be quite clear that the existence of mirror-image twins
is a *genetic paradox*, for it does not make sense that two individuals
who have come out of the same egg and who consequently have
an identical heredity, should differ in the matter of laterality. Scientists have reacted in different ways to this paradox.
Some have set up the hypothesis of an unequal distribution of

chromosomes in the two halves of the egg from which the twins were born; they thus discard the idea of an identical heredity shared by the twins. This theory is generally rejected, since it lacks any sort of experimental confirmation. H. Siemens, who was probably the first to deal with the problem of mirror-image twins, cheerfully discarded the theory of the hereditary nature of left-handedness, despite all the family statistics. Zazzo believes that this paradox calls the whole theory of the twin method in question and considers that we should radically change our views on the relation between heredity and environment. O. Von Verschuer is concerned to save the theory of the twin method at all costs. He believes that there is probably a third kind of twin (besides the monozygotes and dizygotes) and he does not scruple to introduce a hypothetical factor G to explain the mechanism of the genesis of twins. Whatever view we may take of the paradox of mirror-image twins, we are faced here by the right-left problem as something directly connected with the very foundations of genetic research.

Taking Zazzo as our guide, let us look at some of the modern theories of enantiomorphism and of twins in general. Certainly that of H. H. Newman is particularly attractive, possibly because it originates directly from the old right-left myths. The American scientist subordinates the genetic factor to the mechanism of embryogenesis. Taking the well-known case of the Canadian quintuplets as an illustration, he sketches out the plan of an "embryological tree", which enables him to demonstrate in a simple fashion the hierarchy of the dissimilarities and of the enantiomorphic reversals among twins. Newman bases himself on the experiments of Hans Spemann (cf. p. 107). We know from these experiments that the process by which one-egg twins are formed is the same as that which in certain circumstances, can lead to the creation of double-headed monsters, Siamese twins and other monstrosities. We also know, however, that there are only two truly favourable stages in which such formation of twins is possible at all: the blastomere stage after the first cell parting, and a much later one shortly before gastrulation, in which the first organs have already begun to form. In the intermediate stage, i.e. after the second subdivision of the two "totipotent" blastomeres, the operation cannot be successful. According to Newman, the division of the human egg —as in the case of the salamander—can take place at various stages and lead to the formation of various kinds of monozygotic twins, whose degree of similarity is conditioned by the age at which they become separated.

In his opinion the five Dionne sisters came from one and the same egg, but Yvonne and Annette, who resulted from a very early separation, lived for the longest period "together", i.e. in the same half of the egg, resulting from the first twinning division; they are therefore completely identical. Marie and Emilie meanwhile stayed in the company of the fifth sister, Cecile, in the other half of the egg. One day the first half of the egg split open again and, as a result of this new twinning, the two completely identical embryos of Yvonne and Annette came into being. The other half of the egg also divided anew. One half produced Cecile—the "fifth wheel on the coach" while Marie and Emilie continued for a while to live together in the other half. Meanwhile however the somatic organization had become "lateralized". When finally the last pair of twin sisters separated, they were not as similar to each other physically as were Yvonne and Annette; they were *only* enantiomorphic, and for Newman that quality represents a lesser degree of similarity.

According to Newman's theory, twins which have separated late no longer inherit the total organization of the embryo but display the difference which exists between the left and the right sides of the body in a single individual. The idea of a polarity between right and left which is at the bottom of this theory has in Zazzo's view contributed much to its popularity. "Apart from its heuristic value, Newman's theory undoubteldy owes its success to the fact that it is in line with old fantasies and legends about "doubles" and with our age-old dreams about left and right. In the mirror-image twins the duality, the essential twin-ness of every creature, becomes a reality: his right and his left aspect, his social mask and his inner personality (cf. p. 76). But myths are not always a trustworthy guide for the scientist. It is true that they fertilize his imagination and awaken the germs of new theories. But it is also true that they cause him to set out on paths on which there is no confirmation by experience and experiment. Zazzo lists a number of facts which cannot possibly be harmonized with Newman's hypothesis concerning a right-left division which is essential for the creation of twins. If Newman were right, the *situs inversus viscerus*, which in the real world is extremely rare, would appear more frequently among mirror-image twins. Moreover, mirror imaging, which he regards as "reduced similarity" would have to be accompanied by a greater difference in general anthropological character-istics such as the size and weight of the individual, the shape of the head and so on, Yet such a correlation does not in general exist; only in the case of Siamese twins does their mirror image character, as Newman has shown, produce strong differences. (One, for instance, may be a brachycephalic, while the other is a true Aryan dolichocephalic.)

The theory of the Swedish geneticist Gunnar Dahlberg, despite some similarity with Newman, differs from the latter on the decisive point that a twin does not inherit the half organiza-tion, whether right or left, but the whole organization of the original embryo, and in many cases this total organization is from the very first asymmetrical . . . "because of anasymmet-rical organization of the soma brought about by the symmetrical distribution of the genes". Dahlberg infers this from the obser-vation that asymmetrical anomalies—the spots on the wings of a butterfly, one blue and one yellow eye in some Angora cats, or polydactylism in man—can be inherited upon the right *or* left side of the body. In addition to these genuinely genotypical asymmetries, he says, there are facultative or *random asymmetries*, brought about by the variations within the cell itself, and it is noteworthy that these last are supposed to be "primarily" fortuitous. They represent not something which for the moment is unknown, but, on the analogy of the Heisenberg uncertainty principles, something which in principle *cannot* be known. In the case of one-egg twins, both sorts of asymmetries could produce a different laterality, despite the fact that they are genotypically identical. For details of Dahlberg's theory, I must refer you to Zazzo's description and criticism of it. The important thing here is that the structure which was already asymmetrical as a result of heredity is "worked over" by the mechanism of embryogenesis. The division, which is perpendicular to the plane of the laterality organization, gives the twins a different laterality.

According to a third theory, that of D. C. Rife, a further and decisive factor is added to the inherited factor: the position of the twin embryos in the uterus. Where the effect of genes is weak and tends towards ambidexterity, the twins do not possess the laterality which according to their genotype they "ought" to have. The unusual positional conditions bring about the opposite lateralization of the two embryos, of which one becomes right-sided and the other left-sided. Unlike Newman, Rife and Dahlberg seem to make no concessions to the mythical idea which still persists in us of a really essential difference between left and right. Yet, according to Zazzo their theories are founded on false premises, premises which they not only share with each other but also with Newman's theory, which is so opposed to their own. This strange situation merits our attention.

All attempts to explain the mirror image relationship in certain pairs of twins are closely related to statistical findings about the percentage of enantiomorphic twins and the number of left-handed people in various groups of the population. Thus Newman's theory presupposes that there are more enantiomorphic twins among monozygote pairs of twins than among the dizygotes, while Rife, for reasons which can easily be understood, works on exactly opposite assumptions. According to Newman, moreover, there must be more left-handed people among the monozygotes than in any other group of individuals, (dizygotes or non-twins.) Among dizygotes on the other hand, there ought to be no more left-handed people than among

non-twins, and the fact that this is not the case caused him considerable headaches. "This is indeed a real mystery which has
not yet been solved."[8] There is one point, however, on which
Newman, Dahlberg, Rife, and a number of other authors agree:
there are more left-handed people among twins, whether one-
egg or two-egg, than among non-twins, and it is in fact to the
existence of enantiomorphic twins that they ascribe this excess
of left-handed people. *It is through the process of twinning itself, so the
theory goes, that additional left-handed people are produced.*

The assumption that is common to all these writers, as Zazzo
has most convincingly shown, has never been proved. It can
neither be demonstrated that there is any considerable excess of
left-handed people among twins, nor that enantiomorphic pairs
of twins can show a greater number of left-handed individuals
than "artificial" pairs made up of one left-handed and one
right-handed person taken from a population of non-twins.
Zazzo, for his part, points to his own statistics, the fruit of long
years of work in the psychological institute at Paris University.
But he also points to his analysis of the work of the authors he
criticizes. "I for my part", he says, "*have never in all my numerous
researches into the subject of twins, been able to find this excess of left-
handed people of which these authors speak. . . . Between twins and non-
twins there is no difference in the matter of laterality.*"[9] This negative
result demolishes a very important argument in favour of all
those theories which seek to explain mirror-imaging in identical
twins as an effect of the mechanism of twinning itself. Indeed
the whole notion of any special left-sidedness peculiar to twins
disappears, and with it goes the possibility of using this excess of
left-handed people to resolve the paradox arising from the
identity of the twins and the non-identity of their laterality.
Actually for the French psychologist this "loss" is almost a gain.
The wrongly framed question: "Why is mirror-imaging more
frequent in twins than in non-twin pairs—disguises the real
question, which is: "How is it possible that identical twins can
be mirror-images of each other?" Once freed from the illusion
of a mirror-image asymmetry peculiar to twins, the scientific

enquirer can make use of the different lateralities of twins to illuminate the general problem, of heredity.

We cannot at the moment follow Zazzo's bold reasoning any further; let us therefore pause for a little to consider the rather remarkable fact that a number of distinguished research scientists, when engaged on the examination of this problem of the laterality of twins, should have fallen into error and overlooked facts which were quite easy to prove. "The excess of left-handed people and, in correlation therewith, the frequency of mirror-image reversal in populations of twins", says Zazzo, "show themselves to be ideas in which the demands of science, reactions of an emotional character, and strange but familiar images which have nothing whatever to do with science, are closely intermingled. Before it evokes any coherent idea in our minds, the sight of a pair of identical and enantiomorphic twins will shock us as a deliberate and fantastic game on the part of nature. The exact inversion of forms and gestures, *which in the case of others we should never notice at all*, the completely similar hands which are stretched towards us in one case on the right and the other on the left, this complete repetition in two faces which catch us in the crossfire of their glances and smiles, everything serves to produce the impression of an intact mechanism and to create a closed universe in which chance has no place whatever. The shock which we always tend to feel when we come across duplication reaches its climax in the case of such twins. The pair is enclosed in itself and draws us into a magic circle through which, it would appear, nothing can break."[10] And so the research scientist is misled by the myth of the twins, the illusion of the mirror, and this happens all the more easily because statistics can at times produce equivocal results. Rife, for instance, carried out a comparison of the handedness of a large number of twins whom a rich American twin had invited to a party together with their parents. This was before the War; since then, events of this kind have been made into films. But in a group of *children*, whether twins or not, you will always find a bigger proportion of left-handed people than in a group of adults.

6. *Right-man and left-man*

"Are one-egg twins really identical people or is one member of the pair a regenerated right half of the body (right-man) and the other an extension from an original tendency towards the left half of the body (left-man)?" This is a question that many of us have asked ourselves—as did O. von Verschuer[11]—but are we really faced by this alternative? The result of Zazzo's examinations is that "identical" twins are not identical but already display at birth, and even earlier, somatic differences which will become accentuated in the course of their development precisely because of the "twin-situation". He thus releases us from the myth of identity, but does he not free us at the same time from the *myth of the mirror*? Is not the "oppositional similarity" or polarity of twins unmasked as an illusion along with their similarity?

Zazzo shows in his remarkable work how all the facts suggest that the supposed mirror-imaging of the whole individual is never complete and actually involves only an asymmetric reversal of a few specific features, which are themselves already graded features. The delicate gradations of the "laterality formula" themselves make it impossible to regard a slight tendency to left-handedness, to take but one example, as a reversal of a marked right-handedness. I have already observed that *situs inversus* of the internal organs in the case of enantio-morphic twins is very rare, and "that to have one's heart on the left side" is certainly not the opposite of "having one's heart on the right side" (cf. p. 108). So the asymmetries of twins do not exactly correspond, either directly or inversely. In the final analysis, there are real differences which, because of their progressively differentiating effect, turn the twins into two separate individuals. Although Zazzo imputes to them the error of *pars pro toto*, the embryologists have so far been unwilling to abandon completely the idea of the somatic mirror-image. "Right-

man" and "Left-man" live on in the theories of twinning. And if
Gesell is right with his modern version of Plato's Symposium,
even those who are not twins could regard themselves as "left-
men" and "right-men". Following up the work of Bateson, who
sees an aboriginal phenomenon in twinning, Gesell develops the
idea *that every two-sided being can be conceived of morphologically as a
pair of twins.* This means that, contrary to the usual conception
which regards a pair as consisting of two individuals, Gesell sees
a pair in the individual. Through the process of growth, which
never proceeds entirely symmetrically, both sides of the embryo
individualize themselves and in some cases this results in the
formation of twins, which may be said to represent "the extreme
and spectacular case of the phenomenon of hemihypertrophy".
It may happen, however, that one of the twins will perish before
birth and only the other is born. This sole survivor of a twin
pregnancy—and according to Gesell many of us belong to this
category—will then be *a twin without knowing it.* The suggestion
has been put forward that the embryonic formations which have
sometimes been found in human bodies are nothing else than the
unborn under-developed twin brother of a baby. This rather
gruesome theme has actually been treated in literature.

Right-man and Left-man as morphological and functional
mirror-images of each other have sometimes been endowed by
imaginative scientists and philosophers with "diametrically
opposite" *psychic* qualities. The somatic mirror-image would
then be regarded as having its counterpart in the psychological
one. One twin of the enantiomorphic pair, it is held, would also
be psychologically the reverse of the other, and in the case of
non-twins a "left" individual would be the reverse of a "right"
one. We find here once again the whole list of pairs with which
we are already familiar: the "Tree of Sephirot", the spatial
scheme of the graphologist; introversion-extroversion, active-
passive, male-female, and so on; the same qualities which were
attributed to the directions of space are supposed to pertain to
the right-man and the left-man.

If the illusion in the case of the somatic mirror-image reversal consisted in mistaking the part for the whole, then the error is here a double one. On the one hand a spatial scheme is used to express psychological differences, the morphology of the human body provides directly, or by the circuitous route of the brain structure, a psychological typology. The orientation of the body is supposed, on the strength of a naïve psychophysical parallelism, to correspond to the orientation of the soul! Moreover the illusion of the mirror replaces the complicated interweaving of the traits of human character by a rigid static conception of two poles in the psyche, or a double row of qualities which are arranged opposite each other in pairs. "But what a strange mirror it is", Zazzo rightly remarks "which reflects *a woman as a man and the round face of a pyknic person as the long face of a leptosome.*" Zazzo accuses the psychological concept of the mirror of confusing inverted similarity (symmetry) with the confrontation of actual difference. He does not see in the "right-left-development, as does Bouterwek,[12] a simple and efficient natural procedure to enrich the real world by creating essential somatic differences—which also appear to have a spiritual significance. In this kind of thinking Zazzo sees a typical example of the old deceptive systematizing, of that habit of thinking of things "in pairs" which his teacher Henri Wallon examined in the child and which, as he showed, lives on into the scientific constructions of the adult. Causal thinking is here abandoned in favour of analogy thinking.

The discordant character of the typologies which are developed on this basis are too obvious to miss. So the *illusion of the mirror* makes Bouterwek confront the energetic, active, intelligent, virile right-handed man with the womanishly sensitive, passive, emotional left-handed man and the literary talent of the first with the technical and mathematical talent of the second!

In the psychological sphere too, the illusion of the "mirror" has the disastrous effect of disguising something and covering it up, namely the *legitimate* use of the word to describe the process by which every individual forms his own image from the image which is reflected by the other person.

Among those earlier authors who, before depth psychology had ever developed its refined vocabulary, described a right syndrome and a left syndrome, the most important, after Lombroso, is Wilhelm Fliess. According to Fliess, the right side of a person expresses the characteristic psychological traits of his own sex, while the left side expresses the principal governing the opposite sex. "Where left-handedness is present, the character pertaining to the opposite sex seems more pronounced. This sentence is not only invariably correct, but its converse is also true: Where a woman resembles a man, or a man resembles a woman, we find the emphasis on the left side of the body. Once we know this we have the diviner's rod for the discovery of left-handedness. This diagnosis is always correct".[13] For Fliess laterality is directly and visibly connected with the bodily and mental differences between the sexes, yet the polar opposition of right and left does not correspond to the contrast between male and female but to that of normal and abnormal sexuality. Bringing in homosexuality made it possible, at the turn of the century, to discriminate much more strongly against the left, as though the left were somehow connected with the devil. "Since degeneracy consists", so writes Fliess, "in a displacement of the male and female qualities, we can understand why so many left-handed people are involved in prostitution and criminal activities—which are very much the same thing (sic!)— but we also understand how many threads can drag a person down from the artist's way of life to this first stage of dissipation."

A remarkable fact, which shows great honesty of character, is that Freud, although he knew his friend's curious ideas, admitted the possibility that he might himself belong to the "degenerate" left-handed. "In addition", he wrote, "it seemed to me that you regarded me as somewhat left-handed, and if this were the case you would tell me, *for such a revelation of myself would not hurt my feelings.* It is your fault if you do not know all that is concealed within me; the rest of me you have known for a long time. Actually I am not aware of any preference for the left, either now or in childhood. It would be more correct to say that there was a time when I had two left hands".[14] Was Freud then a re-educated left-hander without knowing it? His children deny this. Nevertheless, we cannot fail to be struck by the confession he made in the same letter to Fliess that, like Schiller, Helmholtz and others, he was "right-left blind" (cf. p. 54). "I do not know if other people are immediately aware of their left and right and of the left and right of others; I myself in (early years) always had to think which was my right side, no organic feeling told me."

7. *The left-handed person in a right-handed world*

The left-handed person, as we have seen, has no particular character. He is not a "left" man. Nevertheless, he has his own special problems and presents problems to those around him. If a real sense of inferiority *vis-à-vis* right-handed people is only found in the "pseudo" left handed person, the *social inferiority* of the genuine left-handed person is a fact, and a fact of far-reaching importance. Living in a world of right-handed people, in which everything, from door handles to rules of politeness, depends on the use of the right hand he is compelled to act against his nature unless he wants to meet with the annoyance and pitying astonishment of his fellow creatures. Tools and instruments like scissors, irons and non-electrical coffee grinders are not made for him. If he is a musician he will be well advised to choose neither the violin nor the flute. The table is never "correctly" laid for him; in short, *it is not a practical proposition to be left-handed*. In the army, in particular, his peculiarity attracts attention and he is distinctly unwelcome. This has not always been the case. In antiquity, left-handedness and ambidextrality are often mentioned in connection with the use of weapons (cf. p. 119). They were, if anything, an advantage, in the early stages of military technique, just as they are today in certain forms of sport (boxing, fencing, tennis, ping-pong). "The divine Achilles lifted his spear and Asteropeos, who was ambidextrous (peridexios), hurled both his lances" (Iliad XXI, 163). The Scythians, says Plato, could use the bow alternately with the right hand and the left (Laws). But whereas in single combat the use of both hands, or the left hand alone (a surprise effect!) could be useful, all this is exchanged when the battle is not between two individual opponents but between opposing armies. "If half the men in our army were right-handed and half of them left-handed, the picture would be bewildering indeed", writes G. H. Hildreth. "Efficiency would be impaired and our national defence weakened." So when she speaks of the "right-handed convention" as an important cultural factor, she seems to be thinking primarily of military advantages!

The existence of a minority of left-handed people within a majority of right-handed ones forces human society to make a decision. What is to happen to the left-handed? Should they and can they be turned into right-handed people? Whatever our views on the interaction of heredity and environment, we may take it as proven that, even though we do not come into the world as right- or left-handed persons, the sense of our later functional asymmetry has already been determined at our birth. (cf. p. 118). There is no question here of a bad habit which we can get rid of by saying to the child "Give me the proper hand"; indeed, the child will be entirely in the right if, almost as soon as it has grasped the "difference" between right and left, it answers with the words "My left hand is your right hand". All efforts, therefore, to educate a left-handed person to right-handedness are a form of educational transposition, and can have serious or very serious consequences for the whole person, depending on the degree of laterality involved. Disturbances of speech, disturbances of the emotional equilibrium, impairments of the general character, and indeed pronounced neurotic symptoms, such as a "depersonalization syndrome"—all these things are connected with this kind of forcible re-education—not to mention such things as a harmless tendency to make mistakes in spelling! Conversely, the "re-re-education" to left-handedness of those who have been injured by such forcible attempts to make them right-handed has achieved spectacular results in the way of healing. Hence psychologists and teachers now generally tend to allow the left-handed the maximum of liberty in so far as this is permitted by the laws of the land. Though they are not in principle opposed to re-education, they ask that the education of the left-handed to the use of the right hand should at least take place very early, in the ambilateral stage (before the development of speech), or, failing this, only after most careful tests have been made and the whole developmental history of the child has been taken into account. They also insist that this re-education should be carried out by trained doctors and psychologists and not merely by the family and the school.

Both those who favour re-education and those who oppose it are in agreement on one thing, namely, that the old Platonic dream of ambidexterity has vanished for ever. They stress the necessity of a "good lateralization" which will relieve us of the need to stop and think which hand should be used for this activity or that (cf. p. 114). "As is always the case in biology" says F. Giese,[15] "the illumination of the problem from the developmental point of view shows us that it is not identity but a differentiation which is the sign of progress." The same view is held by Gertrude Hildreth: "Only a genius", she writes, "would have the ability to attain equal mastery of the two hands in a lifetime so as to be able to use them alternately in the same roles." The attempt to help a child to acquire "two right hands" merely results in the acquisition of "two left hands".

Teachers and parents are therefore faced with the task of turning a right-hander into a *good* right-hander and a left-hander into a *good* left-hander. For the most part they do not fully realize that the left-handed child is specially in need of their help. The most important skill, that of writing, would really only come naturally to him if he could write in mirror writing.[16] Left to himself, and endeavouring to produce a right-orientated writing with his left hand, he covers up with his writing hand what he has just written. He strains the muscles of his lower arm and wrist and his fingers are held so that they cannot properly control the finer muscle movements. Fortunately these annoyances can be counteracted by the use of a suitable writing technique and even by means of comparatively simple "tricks", such as holding the paper slantwise. It is, of course, necessary to *learn* this technique: the example of a right-handed teacher is of no use at all. For this reason a few psychologists ask for special classes for left-handed children in the schools, but as far as I am aware, this request has not been complied with in any country— just as all the claims of left-handed people to enjoy equal rights have largely gone unheard. The demand that the left hand be used as freely as the right, and that such use should be fully sanctioned, derived originally from the ideal of ambidexterity: it was, in fact, put forward by the Sophists and by Plato. Rousseau, also, had this now abandoned ideal in view when in his *Emile* he asked that children should not be encouraged to use one hand rather than the other. But in Benjamin Franklin's *Petition of the Left Hand to those entrusted with the supervision of education* we become aware of a desire to help the minority, which has been handicapped by nature, to obtain their rights. The authorities are enjoined to distribute their "solicitude and affection equally to all children". We also come across this essentially democratic point of view in our own times, particularly in the Anglo-Saxon countries, where re-education is rarely attempted. "Democracy", says Kenneth L. Martin, for instance, "is committed to providing an education to every child up to the limit of its capacity. We are not doing our duty when we provide desks with only a right arm, as is so common, especially in our high schools and colleges, lighting systems that make it necessary for a left-handed student to write in his own hand's shadow. Work benches in manual art classes which cater only for the dextral are equally discriminatory."[18] This is quite a widespread problem: should man be adapted to his tools, or the tools to the man? The manner in which a society treats its left-handed people, that ever-present minority, even where the racial stock is entirely homogeneous, is a good indication of the norms and values which are respected in any country. It is, therefore, quite wrong to make fun of the sale of special pens for left-handed people, or of vessels with two spouts.

Valuable as are these modest signs of consideration for the needs of the left-handed, it is of course the progressive mechanization and automation of left which is making things easier for the left-handed in a world of right-handed people. Every electrical domestic or other apparatus, helps him to adapt himself to the society in which he lives. The record player saves him from the misery of having to wind up the gramophone; the electrical tin opener enables him to perform an action which would otherwise be wholly impossible. The typewriter gives so slight an advantage to the right hand that it is scarcely noticeable. Finally, in the next war it will hardly matter whether the general who "presses the button" uses his right hand or left hand for that operation.

Turning to another matter, it is possible that the left-handed person who has been so unjustly treated by Nature might benefit from the revaluation *of left as against right* which now seems to be setting in on political and philosophic grounds. This is of fairly recent origin. In bygone times, when the ideal of ambidexterity held the field, the right cultures were, faced as we have seen, not by cultures of the left but by cultures of the middle and of compromise. It is only comparatively recently that the *left* became the side of *revolution*, simply because it was the side of the weaker, the female side, and also that of the left-handed, who, under the pressure of society revolted against it. Since then, the prestige of the left has been growing steadily. Whereas the left in Parliament was formerly the opposition, the party that had to be content with the "bad" seats, like Judas on the mediaeval miniature, it sometimes has difficulty nowadays in defending the left seats, which have become places of honour, against newcomers. In France, a little time ago, the newly-established *Poujadiste* group would not on any account sit on the right; even the most reactionary of reactionaries feels the need to dress up in revolutionary clothes.

The revaluation of the left cannot, of course, be ascribed solely to the development and modification of the concept of the political left. Is there not, we may well ask, something like a *philosophical left*? In his essay, *Avicenna und die Aristotelische Linke* (1952/1963) (Avicenna and the Aristotelian Left) Ernst Bloch sees the "Aristotelian Left" as moving along a line which does not lead to St. Thomas and to the spirit of the next world but to Giordano Bruno and the flowering of "all-matter". The "left effect" expresses itself in a predominant interest in this world, in a "naturalistic bias" in a "suspension of divine power itself within the active potentiality of matter". Above all, of course, the "leftist" Avicenna's "unity of reason" demonstrates nothing less than the new and gentle sentiment of toleration.

8. *Famous left-handed people*

The defenders of the left have always had a powerful argument in the fact that many important or universally esteemed personalities were notoriously left-handed. Leonardo da Vinci painted alternately with his left hand and his right. His innumerable sketchbooks whose prophetic content has only recently come to be fully appreciated, are filled with that mirror writing which is a characteristic of the left hand. Richard Kobler (cf. note II. 4) includes in his list of "left-inclined" people: Goethe, Nietzsche, Holbein, Adolf Menzel, Frederick the Great, Bismarck, Beethoven, Robert Schumann, Heinrich Heine, Hans Christian Andersen, Uhland, Björnson and Lenbach. The criteria of this somewhat over-imaginative author are indeed of a very general kind—such as the fact that in a number of portraits Heine puts his left elbow on the table. Moreover, he has assembled his gallery of famous left-handed people chiefly to demonstrate a relation between left-handedness and artistic talent. Other authors mention Paganini, Benjamin Franklin, and the anatomist His as being left-handed. Charlie Chaplin, to take an example from the present day, plays the fiddle in a film with his left hand—just like the devil. In another film he counts bank notes just as quickly with his left hand as with his right (counting bank notes and card tricks are actually in high favour as laterality tests (cf. p. 120). A highly characteristic left-hander, who displayed practically all the various aspects of left-handedness, was Lewis Carroll, the author of *Alice in Wonderland*, who was also a teacher of mathematics at Oxford. He wrote mirror writing with great ease, like Leonardo, and the very title of one of his books *Through the Looking Glass* shows how interested he was in the problem of mirror reflection. He stammered, probably as a result of re-education, and did so to such an extent that he had to give up his intention of becoming a clergyman. The dodo in *Alice in Wonderland* is not only an extinct bird but also Lewis Carroll himself, for he had some difficulty in pronouncing his own real name, Dodgson.

VI

THE WORLD OF THEORETICAL SCIENCE

1. *From the cosmos to space—from space to the cosmos*

Right and Left are not experienced, but we carry them—as principles of order—into the "Erlebnisraum". Observation of the plant and animal-world, including man, shows no essential difference between left and right, but only symmetries and asymmetries, neither side having any advantage over the other. But perhaps the world of the scientist, this world which is neither experienced nor observed but "deduced", might impel us to make some kind of a distinction between right and left. Perhaps the "space" of this world can show an anisotropy, "a difference between regions".

The conception of space as something existing independently of the things contained in it, is not self-evidently valid. It is not natural to the child, nor does it appear in the early phases of philosophic thought. Not only was the possibility of empty space disputed for centuries but Aristotle, Plato, Leibniz and Descartes had, one might really say, a metaphysical *horror vacui*; the philosophical need for a "receptacle for all material objects", which seems so self-evident to us, does not appear to have existed for them; things do not necessarily have to be *in something*. If I remove those things one after another in my thought, there does not necessarily have to be something left, like the grin of the Cheshire cat in *Alice in Wonderland* after the cat has disappeared. And so the philosophers and scientists began with the idea of a filled *cosmos*; later—and not without some difficulty— they developed the idea of empty space, the precursor and necessary condition of space; finally, in quite recent times they have come back again to a kind of *cosmos*.

This development from cosmos to space and from space back to a new cosmos has undoubtedly given Aristotle a fresh "epistemological" topicality. Remote as his biological and artefactual analogies may appear to us, and real as must be our reservations when comparisons are made between the dynamic field structure of modern relativist cosmology and that of the "Father of Biology" (though this comparison *can* be made and is most fruitful) it is now much less permissible to throw his views on the structure of the world into the lumber-room of unwanted superstitions than it was in Newton's day, for instance. So Hedwig Conrad-Martius was qutie justified in devoting to him a considerable part of her noteworthy book *Der Raum* (Space). And we are therefore merely following an honourable tradition in beginning with the Stagirite who concerned himself on several occasions with right and left (in *The Heavens*, the *Movement of Animals*, the *Problems* and so on).

Aristotle inveighs against the void into which Atomists like Leucippus and Democritus sought to pack their atoms. For him there are only bodies and their places. The word "space" does not occur once in his *Physics*. Even when it is used, it merely denotes the sum of all the places occupied by the various material bodies, a place being merely an "*accidens*" which means that it does not have the independent existence of a "substance". The bodies, according to Aristotle, form a finite imperishable universe entirely filled by them, which has no beginning and is in the shape of a sphere; outside it there is neither place nor void nor time nor duration. This is the heavenly edifice itself, the "body of the all". Within it the mechanics of natural movement take place. This means that all movements are movements "in themselves", a kind of exchange of positions (Antimetastasis), like the moving of pieces in a Chinese puzzle, in which in some strange way the empty square is missing.

That in this heavenly edifice, or "body of the all", there must be a right and left is for Aristotle a fact attested by experience: the sun and stars—save only the palindromic planets—always begin their movement in the east, i.e. to the right (cf. p. 32). The heavenly edifice has a right and a left *because it is animate*, and, as it were following its example, all other animate things have a right and left too—i.e. all those which have the beginnings of a movement within themselves: man as the "most natural" of beings has them and, in a lesser degree, the animals, particularly the more active ones. Plants on the other hand *have no right and no left*, only an above and a below. Naturally in completely inanimate objects there is an even more marked absence of any difference between left and right or even between above and below. We only speak of such a difference "when referring it to ourselves either as it corresponds to the right-handed parts of our bodies, as in augury, or because of the similarity with our own, as in the parts of a statue, or we thus designate the part lying on opposite sides to our own, i.e. calling right in the object that part which was on our left-hand side . . . in such cases we shall detect no difference in the objects themselves, for if they are turned round, we shall call the opposite points right and left.[1]

In the Aristotelian system, the "more animate" does not necessarily possess the more aboriginal qualities. The Pythagoreans, who also attributed a left and a right to the heavenly edifice, are criticized by Aristotle because they *only* attribute a left and a right to it and not the "more aboriginal" principles of above and below which pertain to every place (*De Coelo* Chap. II. 1.) *The division of movement into left and right does not therefore possess for him the same elemental character as the absolute up and down movement of the part of the elements which seek their "place" either above (fire) or below (earth).* The one *follows* the other in the natural order and is never fully conceived as a *quidditas*. A difference is here discernible between the pair "right-left" and the pair "above-below", a difference which can still be observed in classical physics in that subsisting between vertical and horizontal movement; both Mach and Poincaré have dealt with this subject (cf. pp. 155 et seq.).

In his dialogue on the world systems Galileo criticizes Aristotle for introducing the concepts above and below, which only have meaning in the finished material world, into his general principles. "The cards", he says, "are changed while in our hands."[2] Aristotle was guilty of the same methodological error in regard to left and right, though here the subordination of movement to qualitative change is less obvious. Nobody showed himself better aware of this than Kepler in his elucidations of his own *Mysterium Cosmographicum*: "I am not the first who has been worried by the useless question why the zodiac has been placed at a certain point in space, since it might have been allotted an infinite number of other positions. A similar problem confronted Aristotle: Why, he enquires, do the planets move in a certain direction instead of moving in the opposite one? . . . To which he himself gives the answer, that Nature, among all the various possibilities, always chooses the best and that it was better and more excellent that they should move in a forward rather than in a backward direction . . . a foolish statement, since before movement and bodies existed, there was no difference in sense and direction, there was no forwards or backwards."[3] Thus the "changing of the cards" also refers to forwards or backwards, i.e. to right or left. *Kepler thus anticipates the discussion between Clarke and Leibniz in a most remarkable fashion.* He uses against Aristotle an argument similar to that with which Leibniz sought to convince Newton through the latter's mouthpiece Clarke.

Are we justified in saying—as some people do—that for Aristotle right was "better" than left? There is nothing to suggest that he endowed the right side with any particular ontological, ethical or religious value—let alone that he declared the left to be bad. It is true that we are told that the heavens move towards the right since this is the more excellent side. But, on the other hand, the right is only the more excellent side because the heavens move towards the right. In this connection the observations which Aristotle makes at many points in his works (*Nicomachean Ethics, History of animals, Movement of Animals, Problems*) concerning the uses of the right and left hands are important. Certainly the right hand is the stronger. *But there is nobody who could not develop the ability to use both hands equally. (Nic. Eth.*, V, 10). The right hand appears to be superior because we are used to it, but it is possible through habit to furnish oneself with two right hands. (*Problems*, XXXI, 12.). It is Aristotle's conviction that we could perform equally well with our left hand all tasks for which we use our right. Here we see the *myth of ambidexterity* which dominated the whole of antiquity; left-handedness is not recognized as a biological phenomenon.

For the cosmos to turn into *space* two things are necessary. It must empty itself and it must extend to infinity. These two changes are described in quite exemplary fashion in Cassirer's great work *Das Erkenntnisproblem* (The Problem of Knowledge)[4] and more recently in Max Jammer's *Conception of Space*.[5]

Cassirer emphasizes the central position which the problem of infinity gradually came to occupy in the intellectual life of the Renaissance; more recent writers have perhaps tended to attach greater significance to the problem of emptiness. Empty space in the form of minute interstices or "pores" had insinuated itself into the thought of the first atomists long before the Cosmologues of the Renaissance felt the need of it as a background to their superb new picture of the world, and an even longer time before the drastic demonstration of the Burgomaster of Magdeburg. When Campanella, Gassendi and Giordano Bruno came upon the scene—to name but a few of the great thinkers of that age—space became infinite and limitless. Both developments resulted from the pressure of natural science which was gradually becoming mathematical. The question whether any mental or visual image of space could be formed was of minor importance: indeed endless emptiness in which our material world swims like an island did not appear more difficult to imagine than a finite animate globe, beyond which there is in the strictest sense of the word "nothing". Nor can one say that regions or directions are more difficult to imagine in an endless empty space than in "the body of the all". Lucretius had already drawn the picture of an anisotropic and yet endless space. Precisely because space has a definite direction, namely the vertical one, it was in his view endless, like time. For were it not so, the atoms would long ago have fallen "downwards" to the bottom of space and formed themselves into lumps.

The essential difference between the Aristotelian and the classical conceptions of space does not reside so much in the fact that different qualities are attributed to space or the cosmos, but rather in a complete inversion of the whole conceptual framework. For Aristotle space simply could not be empty because an empty space was irreconcilable—as he strove time and again to show—with a cosmic order of places and things. Now, however, spatial symmetry, i.e. the equivalence of above and below, of back and front, of right and left is *demanded*. This or that must not be the way it is, otherwise the places and directions in space would not be equivalent and interchangeable. Even today our whole scientific thought is dominated by that conception.

Self-evident as the demands of symmetry may be, it is not self-evident that we should comply with them, nor is it easy to do so. It is not enough for the physicist to assure us that his space is the "space of the geometers"—endless, homogeneous and isotropic,[6] in short a neutral medium in which all bodies from the greatest to the smallest move undisturbed according to everlasting laws. It is not enough that he excludes above and below, right and left from his formulae. Despite all his precautions it may turn out that, knowingly or otherwise, he favours a "system of references" and introduces directions into this nominally neutral space—perhaps by the mere act of even permitting the movement of bodies within this infinite emptiness at all. The theory of Relativity criticizes classical physics for its lack of symmetry (cf. pp. 164 and 171). But the discussion about the "difference between the regions in space" had already begun

in Newton's day, a discussion associated in the main with the
four great names of Newton, Leibniz, Kant and Einstein.

2. *The correspondence between Leibniz and Clarke*

A literary fashion of former days delighted to bring together
important men of different ages in some Olympus or Heaven—
or possibly even in Hades—for purposes of instructive discussion,
thus confronting with one another those who had at one time
been unjustly separated by time and distance. Here we must
deny ourselves the pleasure of this charming poetico-philosophi-
cal game. Let us nevertheless try to treat this problem of space,
in the form in which we have set it out for ourselves, without
regard to chronological sequence. I believe that many of the
difficult questions which are bound up with it could be solved if
it were possible to allow Einstein to defend Newton against
Leibniz, or Leibniz to put his case against Kant. The injustice
of history has forbidden this to happen; the only real discussion
was that which took place in the form of exchanges of letters
between Leibniz and the man whom Newton had entrusted
with the defence of his views, the theologian Samuel Clarke.
This ended with Leibniz's death. Newton lived on for several
years—as warden of the Mint—and probably regretted the
fact that he had not suppressed the third (philosophical) volume
of his *Principia*, as had been his original intention. "Philosophy
is such an impertinently litigious lady that a man has as good be
engaged in lawsuits as have to do with her." It is true that he
would not have thus regained peace in his soul, for, deeply
religious man that he was, he was less concerned in this discussion
about space than about God.

   If, despite the fact that the words right and left only occur in a
single passage, we see in the Leibniz-Clarke correspondence a
discussion of the "left-right problem", we shall in doing so be

following Kant. At the end of his treatise *On the first Ground of Distinction between the Regions of Space*, Kant expresses Newton's conception of space as follows: "If we imagine the first thing ever to be created to have been a human hand, then it must of necessity have been either a right or a left one and, in order to produce the one, a different act of the creative cause was necessary from that whereby its counterpart could be made."[7] In contrast to this, according to Kant, "many of the more recent philosophers"—the reference is to Leibniz and Wolff—"assume that space only consists of the external relationship between the parts of matter that exist next to each other", or—to quote Weyl's interpretation—according to Leibniz's view, "it would have made no difference if God had created a "right" hand first rather than a "left" one. One must follow the world's creation a step further before a difference can appear. Had God, rather than making first a left and then a right hand, started with a right hand and then formed another right hand, he would have changed the plan of the universe not in the first but in the second act, by bringing forth a hand which was equally rather than oppositely oriented to the first created specimen."[8]

The reader will seek in vain for this paradigm of the right and left hand in the Leibniz-Clarke correspondence.[9] It was thought up by Kant. The witty Leibniz *might* have conceived it, Newton most certainly could not have done so—still less Clarke; indeed it is questionable whether Newton would have accepted this Kantian interpretation of his theory of space which finds its proofs not in mechanics but in geometry. For all that, Kant gets at the essence of both conceptions of space which are set against one another in this correspondence, though without any hope of a reconciliation between the two. The creation of a right hand is only possible if we assume a Newtonian space, playing the same part with regard to the first act of creation as the one hand in Leibniz plays with regard to the other, namely that of a *reference system*. Newton's space is "absolute". It remains "in its own nature, without relation to anything external . . . always similar and immovable";[10] it has its own reality independently of the existence of all matter. Only into such a *pre-existing* space can a right (or left) hand be created by means of a first creative act (or, stated more precisely, by the first act of creation of a material object). It is not a left or a right hand by reason of the order of its parts with respect to each other, but by reason of their different disposition with regard to absolute space. For Leibniz on the other hand, space—and the same incidentally is true of time—possesses no absolute reality; it cannot exist on its own, that is to say it cannot exist before matter; it is nothing in itself, it is indeed a mere scheme that possesses no more reality than a family tree in which every person has his allotted place. "The author contends that space does not depend upon the situation of bodies. I answer 'tis true, it does not depend upon such or such a situation of bodies; but is that *order*, which renders bodies capable of being situated, and by which they have a situation among themselves when they exist together as time is that order, with respect to their successive positions. But if there were no creatures, space and time would be only in the ideas of God."[11] So the hand that was "first" created is actually neither a right nor a left, and only the second act of creation brought forth a *pair* of hands which had a definable left-right relationship, i.e. furnished a mirror-image of one another without God Himself being able to say whether the one that was first created was a right or left one.

Correctly recognizing the danger by which "his" space was threatened through the empty space of the physicists, Leibniz declares himself unable to discover the emptiness invented by Herr Guericke and shows himself quite content not to do so, for the more matter there is "the more opportunities there are for God to display His wisdom and power".[12] His correspondent, Princess Caroline, hesitatingly draws his attention to the experiments of the Burgomaster of Magdeburg and admits to being impressed by "Mr. Clarke's knowledge and clear ways of reasoning". The philosopher does not answer her and persists in his ostrich-like attitude in regard to empty space. He cannot entertain the idea of a finished material universe perambulating around in an empty space; this would imply *agendo nihil agere*; in this way—and here Leibniz definitely takes up the position of the modern Positivist—no change would be produced which could be observed by anyone. Empty space, we should add, is rightly or wrongly regarded by Leibniz as a preliminary step towards, or a necessary precondition of, Newton's absolute space, and every effort to make space absolute confronts God with an alternative that is unworthy of Him: east or west, that is to say left or right. For were space anything in itself, then the various points in space could be distinguished from one another and we would be faced with what was for Newton the unavoidable but for Leibniz the inadmissible question: "why God, preserving the same situations of bodies among themselves, should have placed them in space after one certain particular manner, and not otherwise; why everything was not placed the quite contrary way, for instance by changing East into West (cf. p. 143). But if space is nothing else, but that order of relation; and is nothing at all without bodies, but the possibility of placing them; then those two states, the one such as it now is, the other supposed to be quite the contrary way, would not at all differ from one another. Their difference, therefore, is only to be found in our chimerical supposition of the reality of space itself. But in truth the one would exactly be the same thing as the other, they being absolutely indiscernible; and consequently there is no room to enquire after a reason of the preference of the one to the other."[13]

Leibniz criticizes Newton for failing to understand that the conception of space developed in his *Principia* compels us to think of points in space as distinct though indistinguishable entities. Newton, he declared, was going against the principle of the identity of *discernabilia* and so against the law of sufficient reason, which governs not only the whole of metaphysics and natural theology but also the dynamical principles of natural science. Nothing happens without a sufficent reason. Even God cannot choose between two possibilities without having a reason for his choice. If therefore you have two things which are equally good and which cannot exist together—such as this world and its mirror-image—the one not being preferred to the other either on its own account or in conjunction with something else, then God will create neither. By subjecting God Himself to the law of sufficient reason Leibniz does not feel that he is in any way detracting from His divinity. That indeed was precisely what his opponent was doing. Newton, he claims, debases God to a clockmaker, who must continually be winding up the world and fiddling around with it, whereas he, Leibniz in allowing God to act according to "the beautiful pre-established order", was holding Him in higher estimation.

Right and left indistinguishable, space nothing but a network of relationships—only if these things are assumed, so says this philosopher, can God be held to act otherwise than arbitrarily. "For the first time in the history of thought", writes André Robinet in his excellent little book on Leibniz, "he tries to see as God sees." Thus in his correspondence with Clarke, Leibniz is not concerned with the validity of Newtonian physics —he had no better physics to set against it, nor for that matter did any other philosopher for the next three hundred years—he was less concerned with the validity of Newtonian physics and with the physical necessity for the Newtonian conception of space than with the theological significance of the famous sentence in the Principia: *Absolute space remains by its very nature without relation to any external object ever the same and immobile.* This sentence he regards as "theologically incorrect", that is to say, not in harmony with the concept of God—a reproach that would necessarily hit Newton all the harder in so far as the latter, as is clear from an often quoted letter to Bentley, was always concerned with principles "as might work with considering Men for the Belief of a Deity". In much the same way as his teacher Isaac Barrow, he regarded his theory of space as a contribution to the knowledge of God and of His attributes. Being no less good a Christian than Leibniz, however much he might oppose him—he was in fact seeking proofs that the world was created by God's free and arbitrary will. He discerned such a proof in the movements of bodies in absolute space. And that is why upon absolute space there hung, as Hedwig Conrad-Martins says, his whole mystically committed heart.[14]

The Leibniz-Clarke correspondence shows that the difference between Newton's and Leibniz's conceptions of the nature of left and right has deep roots. In the final analysis it derives from the difference in their conceptions of God. Although both of the two partners believe in a God who created the world out of nothing, it is not of the same God that they speak. The God of Leibniz is primarily all-knowing, that of Newton "all-mighty". For Leibniz God is—as He is for many modern scientists— essentially the embodiment of the reasoned belief in an inner harmony between "the realm of nature" and the realm of purposes": *Harmonia universalis id est Deus* (Kant has a similar saying "The theological idea serves to enable the whole world complex to be regarded from the standpoint of unity"). Newton however insists—in Kantian terminology—not only on a *world architect* but on a *world-creator*, indeed upon a *Lord*: ". . . a being however perfect, without dominion, cannot be said to be Lord God . . . He is omnipresent, not *virtually* only, but also substantially. . . ." (*Principia* Vol. III, The World System).

In short, Newton avers, according to Clarke, that the "sufficient reason" for the fact that matter turns in one direction rather than in the other, can in certain circumstances be nothing but the mere will of God. Which of the two was right, theologically speaking, we are unable to judge. . . . One's impression is however that Newton showed the deeper religious feeling, though contemporary theologians did not always greet the assistance given by this "outsider" with enthusiasm. (Bishop Berkeley, for instance criticized Newton, because by identifying God with space, he had given Him parts and thus diminished His unity and perfection.) Hermann Weyl holds Leibniz the rationalist to be definitely in error and Clarke to be on the right track, an opinion he bases on philosophical and religious grounds; he believes chance to be an essential feature of the world. Yet he adds that "it would have been more sincere to deny the principle of sufficient reason altogether instead of making God responsible for all that is unreason in the world".[15]

In view of these controversies about absolute space and the possibility of distinguishing right from left we have to ask ourselves whether the idea of the omnipotence of God really implies a particular conception of space or a particular kind of physics such, for instance, as that of Newton. The modern scientist is in the main inclined to answer this question in the negative. He either carefully separates science from religion and faith from thought, or he sees in the ordering of nature a source of the knowledge of God. To seek to establish a connection between scientific propositions and particular "attributes of God" seems to him archaic. Yet we must not forget that in the history of science the tying together of scientific and theological problems had often proved fruitful—particularly in times of crises.[16] Thus Pierre Duhem declares that the theological concept of an omnipotent God has made possible new experiments in thought and so freed our minds from the finite frame in which Greek thought had enclosed the universe. In the "Correspondence", Newton and Leibniz still unreservedly define truth as the thought of God or as insight into the divine creation (whereas Luther had rejected all identification of human with divine thought).[17] Even today at certain turning points the new epistemological situation cannot be better described than as "from the standpoint of God". Such a turning point was quantum physics with Heisenberg's principle of uncertainty. Such thoughts have also become current again quite recently in connection with the discovery of the apparent "distinguishability of left and right", and it is far more than a mere jest when a contemporary physicist speaking of the "non-conservation of parity", and possibly following up Kant's paradigm, declares himself "shocked not so much by the fact that God is left-handed but by the fact that He still appears to be ambidextrous when He expresses Himself strongly" (cf. p. 182). Certainly it was as a result of such "theological" reflections that Leibniz was able to complete what is now considered to be a surprisingly modern analysis of our conceptions of space and time and to reach a relativistic conception of movement. The famous correspondence between Leibniz and Clarke . . ." says H. Reichenbach, "presents us with the same type of discussion which is familiar from the modern discussions of relativity and reads as though Leibniz had taken his arguments from expositions of Einstein's theory".[18]

It is particularly astonishing to the modern reader that Leibniz, "without physics" and most certainly without any knowledge of the experimental findings which of necessity led two hundred years later to the general theory of relativity. *should have recognized the weak spot in Newtonian physics in that very idea of the "Difference of the regions in Space" which Kant so readily accepted.* The inherent difficulties in Newton's theory—and there are a number of them[19]—can actually be formulated in "termini of direction", as differences of direction which have been introduced into, or discovered in, space. Leibniz's criticism that Newton, by introducing absolute space, had accorded preferential treatment to one side, namely to that which had been realized in the creation, has its physical meaning.

To understand why in Newton's world the directions are not equivalent one need not penetrate very deeply into the secrets of the theory of relativity. It is enough to study from this viewpoint any of the numerous popular expositions, e.g. Lincoln Barnett's *Einstein and the Universe* and certain epistemologically orientated writings of the pre-relativity epoch.

In the first place, there is here an almost undisguised *difference between vertical and horizontal movement* which finds expression in the fact that Newton's second axiom—that the force necessary for the acceleration of a body must be proportional to its mass— does not appear to hold good in the case of a body in free fall. Ever since Galileo carried out his experiments with falling objects from the leaning tower of Pisa, we have had to accept the fact, for better or worse, that the speed of fall of all objects, great and small, heavy and light, varies only as a result of the resistance of the air. In a vacuum a cannon-ball falls no faster

than a handkerchief. On the other hand objects hurled horizontally with equal force attain even in a vacuum a speed determined
by their mass and other characteristic properties. "Does this not
seem", asks Lincoln Barnett, "as though the factor of inertia was
only effective horizontally?"[20] To Newton, too, such a distinction
between vertical and horizontal direction must have appeared
unacceptable. The law of gravity gets rid of it only by means of
a kind of trick, by saying that gravity, the mysterious attraction
that operates between bodies, increases exactly like inertia with
the mass of the attracted body and so counteracts it in exactly
the right measure and compensates for it. Inert and heavy mass
are the same. Nevertheless Poincaré always thought it strange
that a ball rolling for a very long time on a marble table is
supposed not to be subject to the attraction of the earth.
"Teachers of mechanics", he adds, "usually pass rapidly over
the example of the ball."[21] This strange agreement between
gravitation and inertia was more or less uncritically accepted for
some two hundred years—until Einstein hit on the idea that it
could not be the result of mere chance. The principle of equivalence in the general theory of relativity, illustrated by the well-
known experiment in thought of the lift in the cosmos, turns the
practical indistinguishability between inertia and weight into
one of principle; it sees in the inert and the heavy mass not two
distinct things which happen always to have the same value, for
it affirms that there is no way in which a movement caused by the
forces of inertia can be distinguished from one caused by gravitation.[22]

The second paradox of classical mechanics which is concerned
with "directions" is to be found in the difference between a
movement which maintains its direction and one which is
continually changing it. The circular movement, which already
appeared to the men of antiquity to be the most perfect, seems
"from its very nature" to be different from the straight-line
movement and has, as Huyghens says, "its criterion which the
straight-line one does not possess". Only the latter is equated
with rest by the classical principle of relativity, for if this

principle is to be valid, bodies in movement should change neither their speed *nor their direction*. Like the first paradox, this second one soon shocked some of the more critical spirits. "Why", asks Poincaré, "is the principle (of relative motion) only true if the motion of the movable axes is uniform and in a straight line? It seems that it should be imposed upon us with the same force if the motion is accelerated, or at any rate if it reduces to a uniform rotation. In these two cases, in fact, the principle is not true."[23] For, as he says elsewhere, "if the sky were for ever covered with clouds, and if we had no means of observing the stars, we might nevertheless conclude that the earth turns round. We should be warned of this fact by the flattening at the poles, or by the experiment of Foucault's pendulum." As we know, the principle of equivalence of the general theory of relativity did away with this second paradox along with the first. The new ideas of space and time thus showed their fundamental superiority independently of their subsequent experimental confirmation.

Posterity has to some extent been severe with the discoverer of "absolute space". Newton, who usually carefully separated religious from metaphysical problems, makes—so Jammer remarks—a single exception to this practice in regard to space. According to Hermann Weyl "Newton's picture of the Cosmos is obviously a more limited one than that which a hundred years previously had filled the passionate soul of Giordano Bruno". He criticizes him in his *50 Jahre Relativitätstheorie* (50 years of the theory of relativity) because in his system he cannot distinguish between uniform translation and rest, as had been his original intention. "In order to achieve his end, Newton takes refuge in a cosmological hypothesis and a substitution of concepts which stand out oddly in the magnificent and firmly founded edifice of his *Principia*."[24] Curiously enough, no modern thinker has displayed a better appreciation of Newton than the destroyer of absolute space and creator of the theory of Relativity—Albert Einstein. The well-known "Newton, forgive me!" in Einstein's autobiography was much more than mere courtesy towards an opponent, for he continues, "You found the only way which, in your day and age, was just possible for a man of the highest mental and creative power".[25] Newton's great achievement was that he introduced absolute space as the "independent cause of the inertial behaviour of bodies"—i.e. he recognized "that the purely geometrical entities . . . and their development in time do not fully characterize movements in their physical aspects". However "ugly" we may find the existence of an infinite number of inertial frames in preference to other rigid frames, such a separation of kinematics and dynamics was undoubtedly necessary to give expression to the laws which were known at that time and, this separation not only shows a weak spot in Newton's system but also his *wisdom*. Newton himself was better aware of the weaknesses inherent in his intellectual edifice than the generations of learned scientists which followed him, a circumstance which, so Einstein tells us, always aroused his deep admiration.

When we think of this "Einsteinian" Newton, we can only regret that he let himself be defended in the Correspondence by Clarke. Newton the *physicist* might well have made a better showing against the quick-witted Leibniz than Newton the *theologian*. He would have shown that it was *physical* grounds which led him to introduce absolute space and that—contrary to the criticisms later levelled against him—throughout the greater part of his career he had "left God outside the door of his laboratory", as we still expect every genuine scientist to do. Above all, he would have been able to show that, *because* he was a physicist, he did all he could to produce adequate reasons to explain the differences between the regions in space, and that he did this because in introducing absolute space it was in no way his intention to confront God with the alternatives of having to choose between left and right. To a theologian like Clarke it might be quite acceptable to burden God with such decisions. It would not be acceptable to a physicist, and that was why all physicists welcomed it when Einstein actually saved God the trouble.

3. *Kant was wrong here*

Nobody read the correspondence between Leibniz and Clarke more attentively than did Immanual Kant. Cassirer tells us that the knowledge which he gained from it contributed much towards enabling him one day to cry out. . . *The year 69 showed me a great light.* Through Kant, the correspondence gains that additional point that makes it so interesting for modern research into knowledge, for while Leibniz and Newton necessarily talk past each other, it almost seems to us today as though Leibiz had been attacking Kant in advance.

On the first Ground of Distinction between the Regions in Space, in which Kant demonstrates Leibniz's and Newton's conception of space by the paradigm of the right and left hands, was written shortly before the "light-bringing" year. For this reason it is by many regarded as unimportant[26]—Gottfried Martin speaks of Kant's transient liking for Newton—but the difference between left and right is carried by Kant from the pre-Critique period into the post-Critique period. (*Dissertation of 1770, Prolegomena, Metaphysische Anfangsgründe der Naturwissenschaft* etc.) The paradox of symmetrical figures decisively determines his conception of mathematics and of space which—transferred also to time—becomes a supporting pillar of his philosophy.

The difference between left and right had fascinated Kant from the start—even as it fascinated Mach and Pasteur—and it did so not only in reference to the nature of space: *Even however as the right side appears to have the advantage of mobility over the left the left has over the right in the matter of sensibility.*[27] Also in *Die Macht des Gefuhls* (The Power of Feeling) he seriously enquired "why soldiers always start off with the left foot". The great riddle for him however was the existence of three-dimensional symmetrical objects such as opposite spherical triangles, right- and left-handed screws and snail shells, the two sides of the human body and especially the two hands, ears etc. These shapes, apparently identical in all their parts, cannot be brought into coincidence. Yet it is precisely in their "congruence", according to Kant, that their geometrical identity consists, and he cannot help feeling it paradoxical that objects which are similar in all their parts are nevertheless not wholly similar. Does not this mean that an object such as a "right hand" cannot be completely described by the relative position of its parts, and that something must be added to its intrinsic properties in order to distinguish a right hand as such.

In the treatise *On the first Ground of Distinction between the Regions in Space*, Kant identifies this "thing that must be added" as the relation of the object in question to space, an absolute, oriented space—allegedly Newton's space, though it is very doubtful whether Newton would have accepted this help from Kant. The purpose of the treatise is to show that not only the facts of mechanics but also those of geometry provide conclusive proof of the truth of Newton's assumptions. Yet in the *Prolegomena*, during a period that is to say in which his own theory of space was completed, Kant makes use of the same paradox of right-left symmetrical objects to defend a thesis which is almost the opposite of that of absolute space: namely the ideality of space, "his" space of the transcendental aesthetic, which is an *a priori* form of pure perception. That the difference between two symmetrical figures cannot be explained by the arrangement of the parts, and that—in his belief—it is not at all of a logical conceptual kind, now proves to him the intuitive character of geometrical figures and of space itself. Both these have only one thing in common: they are directed against Leibniz. One of them is intended to disprove the relativity of Leibnizian space, the other its intellectuality. It is not hard to understand that Louis Couturat should find the double use of the same reasoning suspicious; it suggests that in both cases, or at any rate in one of them, the argument is inconclusive.[28]

For the mathematician, Kant's error is to be found in his assertion that enantiomorphic objects cannot be explained in logical conceptual terms by a change of sides in the arrangement of their parts. In his *Philosophy of Mathematics and Natural Science*, Weyl shows that the difference is entirely logical-mathematical and indeed of a combinatorial nature. A permutation of given linear independent vectors determines the sense of the rotation. While the philosopher seeks the solution of the right-left riddle in transcendental idealism, the mathematician finds it in the distinction of odd and even permutations. It is also possible to gain an understanding of this by means of elementary geometrical concepts. In his book *Raum und Zahl* (Space and Number)—in the chapter entitled "On the distinctions between the Regions of Space"—Kurt Reidemeister uses a very simple logical process to deduce the existence of enantiomorphic objects from the axioms of Euclidean geometry, which for their part can be formulated through statements regarding the distances between points.[29] Assuming that these had furnished us with two figures which corresponded in the positional relationships between their parts without our knowing whether they coincided with one another or were symmetrical in reflection, their congruence or incongruence will then be determined not by the positional relationships of the parts of each of the two figures taken by itself, but by the positional relationship of the parts of the composite figure which is made up from both of them and to which the relation of incongruence applies. Thus a figure does not acquire the features that distinguish it from its mirror-image by being situated in space. The property which makes a hand a *right* hand does not derive from its relation to space but from its relation to the other hand, in this case the left hand.

Since the modern logic of relations has replaced that of subject-predicate, which goes back to Aristotle, it is naturally easier for us to describe the circumstances which obtain here. It would seem that Kant had what would even in his own day have been considered an erroneous conception of the logical and ontological status of relations and of their place in mathematics. In his opinion, mathematics was the science of magnitudes and geometry was concerned with the measuring of distances, angles, etc. The correspondence of the position of the parts of similar figures and their mirror-images was therefore understood by him as denoting the fact that the points of one figure can be completely identified with the points of the other, i.e. that the corresponding pairs of points are equidistant. But in mathematics there is not only a magnitude relation between the parts of a figure; there is also one of order. If one has in mind the relations of order, one naturally cannot say of enantiomorphic figures that they are identical in regard to the position of their parts; the "difference" between them is expressed precisely by the concept of reflection, a concept that has nothing to do with perception but signifies a well-defined logical operation.

Kant knew of Leibniz's *Analysis situs* when he wrote his trentise *On the first Ground of the distinction between the Regions in Space*, but he suspected—wrongly as we now know—that it had never been more than just a "notion", and he had quite a wrong conception of it. "I do not know to what extent the object which I here place in front of myself for contemplation is related to that which the great man had in mind. To judge by the sense of the words, however, I am here seeking philosophically for the mathematical premise which he used to establish the magnitude relationship."[30]

One sees that Kant underestimated the intentions of the "great man in question". The *analysis situs* is much more than a "notion". It is a much more ambitious undertaking. By setting itself the task of defining in terms of pure logic certain concepts which till then had been left to intuition, it tears them away from traditional philosophy and claims them for mathematical logic. This means that it is sufficient unto itself and does not need any philosophy which would have to postulate the existence of something which already possesses the same existence as all other mathematical objects.

Kant's ideas about mathematics and especially about geometry have been the subject of long discussions which have not ended even today. People have attempted with varying degrees of success, to harmonize his "pure intuition" with the actual development of mathematics. There was thus created the image of a rather "Leibnizian" Kant who, according to some people, was even demanding a geometry with several dimensions which would be the "true" one, and hence thought that the geometry of Euclid was not the only possible one that was free of contradictions (cf note, 13 to Chapter III). But the example which he was continually adducing, of the right-left symmetrical object and his interpretation, which was that of one who was wholly dependent on perception, shows the size of the unbridgeable gap that separates him from the present-day mathematician. The latter does not know of any such intuition which might permit him to absorb directly the substance of geometrical propositions and hence—as Reidemeister points out—would really make his own work superfluous. The essence of any geometry, including that of Euclid lies for him in its conceptual structure, in the logical derivation from the axioms of a system of propositions free of all contradictions. It has nothing to do with the existence of things or with their properties. Above all it has nothing to do with space in itself.

Thus it is Leibniz and not Kant who is the father of modern geometry, which, logically regarded, cannot be distinguished from any other mathematical discipline, say from the theory of numbers or from the calculus of variations. It is also not Kant but Leibniz who is the originator of that quite different kind of geometry which came into being in conjunction with the theory of relativity—geometry as a branch of physics. Reidemeister's noteworthy analysis shows that Kant's "reinterpretation of our way of conceiving space and turning it into an intuition which formed the basis of knowledge . . . eliminates things as bridges to space". He thus blocks one of the new roads of physics. After the mathematician, the physicist too is relieved of a part of his task— the part that consists of answering the question: which particular geometry is valid for physical space. This question, which for a Kantian is meaningless, both on practical and theoretical grounds, has acquired meaning precisely *because* the deductive geometrics are not concerned with material things or their properties. In particular the question about the differences of directions in space which cannot be meaningfully posed either in regard to Newton's absolute space or Kant's aprioristic one, becomes one which can to some extent be empirically determined. Today we can ask whether our universe is really isotropic, whether it is really homogeneous. Or—to put the matter differently—whether it is possible to construct a satisfactory physical theory which reconciles our assumption of isotropy and homogeneity with the observed facts? The sense in which this question can be asked and the—rather surprising—way in which it can be answered will be revealed if we take a glance at modern theoretical physics and its changed attitude towards the question of symmetry.

4. *The principles of symmetry in physics*

The modern physicist enjoys much greater freedom in the choice of his principles than did his colleague in classical times. He can even to a certain extent "choose" his geometry now that the theory of relativity offers him the possibility of looking upon physical quantities as the begetters of geometrical properties. It is however understandable that while he is comparatively ready to sacrifice certain principles, it will need a particularly strong "pressure of experience" to make him abandon others, which appear to occupy a privileged position in the hierarchy.

He clung for a long time to the perceptual space of Euclid, he continued to prize simplicity, and he made the sacrifices that the theory of relativity demanded of him only to save the principle of a consistent causal order.[31] The symmetry of space, its homogeneity and isotropy, was accounted so fundamental and indispensable that it could never be questioned and survived all revolutions. It is in line with this that, as L. Infeld has pointed out, relativistic cosmology has clung to the assumption of isotropy and homogeneity if for no other reason, than because these "assumptions are the simplest possible and they *are not contradicted by experiment*."[32] In Einstein's works there is the implicit assumption that in a proper co-ordinate an observer looking in different directions will never notice that any one of them is preferred. And when, as a result of the discovery of the red shift, the original static Einstein universe breaks down, it is significant that De Sitter prefers to introduce a fifth dimension rather than sacrifice homogeneity and isotropy. He erects a spherical four-dimensional universe embedded in a five-dimensional space in which the demands of symmetry are automatically fulfilled.

By clinging to such demands of symmetry, relativist physics reveals itself to be the direct continuation and crown of Newtonian physics. From its very beginning, natural science has been governed by the idea of symmetry, a symmetry which at first was wholly the visible symmetry of the outer form and only gradually took on the more hidden and sophisticated character that it has today. Up to the time of Copernicus, and even later, only statics, the science of the *equilibrium* of bodies, could be tackled on a scientific basis. According to the ancients, Nature, like the wise, loved rest. But modern science too, in its early stages, was dominated by the principle of balance, to which all movement, according to Galileo, must be reduced, or by the idea of Buridan's ass, who remains immobile between his two bundles of hay. Hence symmetry is only recognized in the state of rest, as in the equilibrium of a pair of scales. It is regarded as destroyed when one of two forces which are in equilibrium becomes stronger than the other and the system sets itself in motion. According to P. Renaud, Copernicus restored symmetry by introducing inertial force and so laid the foundation of dynamics. From this point on, time begins to occur in the equations, and the way is open for a scientific description of movement. "This restoration of symmetry results in the introduction of a new form of energy, kinetic energy, which in the balance represents that part which is absorbed by the forces of inertia and restored by them. In the case of friction, however, the symmetrical pattern of this material point is disturbed again. If one still wants to set up equations, symmetry must be restored by forces of a new kind: the forces of friction through which at least one new form of energy is introduced."[33]

This manner of writing the history of mechanics is of very great interest, for in a certain sense the whole history of mathematical natural science can be regarded as a sequence of such "restorations of symmetry". I should like here to mention only one particularly fine example of this—because it is concerned directly with right-left symmetry and is also analysed in Weyl's book.

We all feel a certain amount of disquiet when we learn for the first time in school of the Oersted experiment. In this experiment a magnetic needle suspended parallel to a wire through which an electric current is sent, is deflected either to the right or to the left. Just like young Ernest Mach, who is quoted here by Weyl, we cannot help wondering how it can be that what appears to be a symmetrical experimental arrangement permits the magnetic needle to behave in this way when logically it should "hesitate" between the two directions and, like Buridan's ass, refrain from all movement. The explanation of this paradox is to be found in the "diverse natures" of electricity and of magnetism. Whereas the electric current, when reflected in a plane laid through the wire and the magnetic needle, keeps its direction, the magnetic north and south poles are interchanged, due to the essential similarity of positive and negative magnetism; the magnetism of the needle has its origin in molecular electric currents circulating around the direction of the needle and changing their direction when reflected by the plane.

Historians differ about the extent to which chance played a part in Oersted's discovery and about whether the Danish physicist himself was aware of the far-reaching effects of his discovery; but whether or not the relations between electricity and magnetism were really discovered in this way, the pattern is very characteristic of the process of discovery and they could have been discovered in this way. Here we have a model of the physicist's way of thinking. He observes a difference between right and left. "This cannot be", he argues, "because the world is symmetrical." And he seeks by the introduction of new factors and forces to restore the symmetry that has been lost.

The usefulness of the symmetry principle in this form has its limits, however. As the activity of the physicist consists in a progressive "ironing out" of asymmetry his creative imagination is indeed set in motion by every observed difference between right and left. Yet this very symmetry, if it is too much in evidence, may be a source of error for him. We are thus concerned with the *illusion of the mirror*, a simulated simplicity comparable to that which we have observed in biology. Misled by optical symmetry, the physicist, on the strength of known principles which he regards as irrefutable, seeks for a hidden "parameter", for a factor which, if added, could restore the "optical equilibrium". Mach deduced, for instance, from the Newtonian "Paradoxes" (cf. p. 154) the existence of some distant heavenly body. But this method does not always prove adequate; for sometimes the really important thing is to search *not in front of the mirror but within it*, not to add something to what is already there but to "duplicate it" and to recognize it in its essential complexity. "There are concepts which for the moment are still simple, but one should perhaps have the courage to foresee that they will become more complicated.[34]

Gaston Bachelard, whom I quote here, has coined the phrase *conduite de miroir*, intending this to be analogous to the expression *conduite du panier* used by anthropologists to describe that stage of emergent man when, in distinction from the animals, he began to use a basket to gather things. The mirror-stage, however, is not just a pre-scientific stage which precedes the scientific stage as gathering precedes agriculture. The law of reflection, which we treat as natural, gives the scientist a workable scheme for all processes in which something bounces off something else. For instance, it provides him with an explanation of the elastic collision (just as—in reverse—the elastic collision explains the reflection of light through the bouncing back of little balls of light). That is why Kepler insisted that all natural phenomena should be reduced to the principle of light, and we can understand how it is that even the modern physicist can be seduced by the intuitive clarity of the mirror law. For there is a seduction here: "The principle of sufficient reason", says Bachelard, "is clearly associated with the law of reflection. At one stroke it associates mathematical law with actual experience. Thus there comes into being at the very root of science a fine type of *privileged experiments* possessing important explanatory powers and themselves fully explained. An event in the physical world is thus raised—quite improperly, let us add—to the status of a medium of thought,[35] of a category of scientific reasoning. This event provokes a lightning geometrization, which should arouse the suspicions of the physicist who is accustomed to the complexity of mathematical physics".[36]

Actually Bachelard is able to show by the example of the ancient and much discussed problem "Why is the sky blue?" how the privileged intuition of the reflection of light can here lead to illusion. Lord Raleigh's formula of 1897, in which the molecule is treated as an obstacle which simply reflects the light, actually conceals the fact that diffused light consists of greater and smaller frequencies than incident light, an important discovery made by the Indian physicist Raman as recently as 1928. "The scientific consequences of the discovery of the Raman effect are well known", says Bachelard, "but can we overlook its metaphysical ones?" In the spectrum emitted by the molecule the French philosopher sees a "number-spectrum which communicates to us the new mathematics of a new world". The azure blue of the sky is therefore for him "as instructive for the new scientific spirit as some centuries ago were the starry heavens above".[37] Bachelard adduces the so-called Compton effect as a second example of the insufficiency of the mirror scheme. We need not concern ourselves with that more closely however, since the discovery of the non-conservation of parity will provide us with another—and better—opportunity of showing how this scheme conceals an "essential complexity".

At the turn of the century Pierre Curie, the husband of Marie Curie and co-discoverer of radium, made an important contribution to the necessary correction of the mirror scheme in his various works on the principle of symmetry. Pierre Curie is one of those scientists for whom experiment is essentially a predictable and correspondingly organized confirmation of purely theoretical reasoning. It was merely on the basis of his theoretical reasoning about symmetry that he discovered Piezo-electricity, "much as the astronomer Leverrier deduced the existence of the planet Neptune on the strength of his calculations without ever having actually seen it through his telescope". The logical analysis of his own discovery showed him "the necessity of a universalization of the laws of symmetry with the object of applying them to all the states of space produced by physical agents", and led him to formulate a number of principles governing the relation between the symmetry of cause and the symmetry of effects. One of these declares that whereas the dissymmetry of effect must always be rediscovered in the dissymmetry of causes, *the effects can nevertheless be more symmetrical than the causes*. We know today that Pasteur and other famous scientists would have been able to avoid certain mistakes had they been acquainted with this principle (cf. p. 95). It was Pierre Curie who coined the phrase: "It is dis-symmetry that produces the phenomenon". "He taught the physicists", says A. Kastler in his jubilee article, "to recognize the true symmetry of a physical phenomenon and the conditions under which a phenomenon can or cannot occur. But for him they would hardly have taken into consideration the fact that the symmetry of a magnetic field is very different from that of an electrical field, that E is a polar vector, which can be represented by an arrow, whereas H is an axial vector with the symmetry of a revolving cylinder"[38] (cf. p. 165).

This was probably the first time in the history of epistemology that anyone had revealed the real meaning of the concept of disturbance in the process of scientific thinking—as a deviation from symmetry which must be regarded not as the "degeneration" of a harmonious equilibrium but as the *source and cause of new phenomena*. It is in the elimination of the old concept of disturbance that Bachelard sees the veritable "Copernican revolution" of empiricism in our time. "We should no longer speak of simple laws which are disturbed, but of complex organic laws which sometimes display certain viscosities and nebulosities; . . . in the delicacy of the ultimate approximations, knowledge receives both its crown and its true structure."[39]

Curie's laws of symmetry take us a long way from the right-left symmetry of two candlesticks on a mantelpiece, a long way from the more complicated symmetries of classical physics, and enable us to grasp the symmetry of motions. The symmetry which here comes to light embraces our progressing knowledge, and at the same time an entire reality in process of creation. We are thus led quite naturally to our modern conception of the scientific process, which is wholly built up on the concepts of the group and of invariance. If the physicist of today has learned "to calculate with an imperfect calculating machine"—i.e. to extract the complete from the fragmentary and guess the reality from mere compossibilities, if his knowledge increases from day to day, despite—or perhaps because of—the fact that, like Alice in Wonderland, he plays his game of croquet with hedgehogs for balls and flamingos for hoops, then he owes this not least to the changed conception of symmetry. His perspective has undergone a complete change; symmetry is something hidden which must be brought out into the open—by experimenting with the groups which may provide him with invariants. The permanent element will be provided by a multiplicity of transformations after the manner of a Leibnizian nomad whose permanence consists in the infinite multiplicity of perspectives. It is thus that the theory of relativity constructs the objective world out of the changes which the phenomena show to different observers. It is thus that quantum physics achieves an acceptable reality through reciprocity and relativity. The criterion of objectivity resides "within the structure itself". It is no longer ideal qualities such as finiteness or a definite metric or isotropy that prove it to be present. And conversely because they are no longer the necessary conditions for a necessarily accepted, objective world, these ideal qualities are no longer indispensble and we can at least consider doing without one or other of them.

Does this hold good even of the right-left symmetry of space or of the four dimensional space-time which in relativist physics has replaced Euclidean space?—Only the child is astonished to see its own image in the mirror. For the adult it would be rather an uncanny experience if at some time or in some place this were not so. But in the world of the physicist sentimental considerations carry no weight. His spirit is prepared to accept the possibility of subjecting even so basic a concept as that of isotropy to experimental control. And if it were then to appear that there are phenomena which undergo a change through being mirrored, he would—however reluctantly—accept the conclusion that space "contains a screw". In an essay on relativist cosmology to which reference has already been made Leopold Infeld stated that the assumptions of isotropy and homogeneity had hitherto held their own. But "it is not at all sure whether further observations, made with new and more powerful telescopes will not force us to change them".

These "further observations" have since been made, but it was not more powerful telescopes that shattered our picture of the world. *The revolution took place in the micro-world.*

5. *Symmetry, invariance, conservation laws*

If we are to form a picture of the nature and importance of this upheaval which we call the non-conservation of parity, a few remarks about the relation between symmetry, invariance and the so-called conservation laws, will be necessary.

Regarded in a wider sense, and divorced from its purely geometrical aspects, symmetry denotes the invariance of a formation to certain groups of transformations. . . To put the matter another way, symmetry according to the modern conception means that a description remains unaltered by any change in the reference system (e.g. displacement of the co-ordinate system, rotation, time reversal, etc.). Incidentally, this reference system need not necessarily be of the space-time kind in the classical sense; it can also be a "multi-dimensional configuration space", and there are also symmetries, which do not concern space or time at all, such as the irrelevance of the zero point of an electric potential.

This identification of symmetry and invariance is what really enables us to understand the threefold role played by symmetry. In the first place, the existence of symmetries is in accord with our daily experience. Secondly it expresses not only an emotional but an intellectual demand, without which there might perhaps never have been such a thing as exact science. For the remarkable thing about scientific thought is that it always starts from something that is permanent and invariant, even though the invariance may later prove not to be absolutely valid! Thirdly, symmetry in the sense of invariance becomes the expression of general conformity to laws and an indication of deep inter-connections which disclose the actual structure of the world and its intrinsic physical nature.

Unconsciously, or in the form of the principle of sufficient reason, it has guided the steps of the scientist from the very beginning; isotropy and homogeneity are concepts that go back to the very beginnings of scientific thought, and the more hidden invariance of the laws of nature was also recognized at an early stage under a "Galilean transformation of co-ordinates". But the ontological value of the invariance principles was barely visible before the coming of the theory of relativity which, after all is "only another aspect of symmetry". According to Wigner[40] Einstein was the first to recognize the significance and general validity of the invariance principles. The change in our habits of thought which took place on this occasion lay in the fact that we ceased to be satisfied with establishing the symmetry of the laws which we had empirically discovered; instead, by using the "symmetry-group" of the system under examination we derive new laws which have hitherto been hidden from us. Amongst others C. N. Yang, a physicist of whom I shall have much more to say, stresses the fact that in the theory of relativity the connection between the laws of symmetry and the dynamic laws of physics takes on a much more integrated and inter-dependent relationship than hitherto. Above all else, the realm of physics now also includes invariances which not only have nothing to do with our everyday experience but actually run counter to it and must therefore be revealed by complicated experiments.

The theory of relativity was only able to survive the rather difficult transition to the quantum theory standpoint because those laws of physics which express a fundamental invariance or symmetry appear essential to us today. Actually in quantum physics the study of symmetry has shown itself to be genuinely creative. Dirac's quantum mechanics leads us to predict the formation and annihilation of pairs of particles bearing opposite charges, and so to the concept of anti-particles and their successive discovery—if this word can be used to describe the new method which consists in provoking the appearance of a physical object "through the multiplication of the repeatable interactions in which it can be involved". Positrons, antiprotons, etc. are theoretically anticipated as a consequence of the in-variance of the physical laws to Lorentz transformations. From among a number of possible interpretations one chooses the most symmetrical—and here, rather amusingly, one theory might actually be reproached for its suspicious excess of symmetry!

It is characteristic of quantum mechanics that in it symmetries, or invariances, find expression primarily in the *conservation laws*. This is the name given to certain basic laws of physics, in accordance with which certain physical entities can neither appear nor disappear. The important part played in classical physics by the conservation laws in regared to energy (energy principle) momentum (inertia principle) and electrical charge is well known. But it was only in our own century that the relationship between conservation laws and symmetries was fully recognized.

In general the symmetries and invariances which appear in the laws of physics imply conservation laws and are so to speak identical with them mathematically. Thus from the invariance of nature to rotations in space (isotropy of space) there follows the conservation of the angular momentum. Similarly there follow from the homogeneity of space and time the conservation laws for momentum and energy which merge in the theory of relativity. An exception is formed in classical physics by the discrete symmetries, i.e. symmetries where the changes which are undergone by the physical system, and which turn out to be irrelevant, are not arbitrarily small. In elementary particle physics, where the distinction between discrete and continuous symmetries disappears, these symmetries nevertheless do lead to conservation laws—"non-classical" ones of course—which concern concepts like strangeness[41] and parity that have no meaning in classical physics. In contrast to the theory of relativity, which has actually combined two conservation laws to form one, quantum physics has the tendency to multiply the number of conservation laws. In the sub-atomic world in which causality, as bound up with the permanence of the object, is now of less importance, these conservation laws still guarantee the law-abiding nature and character of the particle as a well-defined physical entity. As de Broglie says, "When the particle escapes from its space-time frame", when, to use an expression of Bohr's it transcends this frame, the law abiding character expressed through the conservation of energy and momentum still has a meaning. Moreover, the conservation laws provide us with selection rules which tell us whether a particular reaction satisfies their requirements. So they considerably restrict the steadily increasing number of particles discovered since 1953—there is a saying:

"To every physicist his particle!"—telling us which are possible and which are not. Yet their functioning affords no knowledge of the details of the process concerned or of the complete dynamic behaviour of the system. And once a particle has become possible—or rather has been *made* possible—then the chance arises of actually encountering it on one of the thousands of photographs which are analysed every day in nuclear research establishments.

The validity of the conservation laws is an object of research, and can be tested. In McGraw Hill's *Encyclopedia of Science and Technology* (1960) we find an instructive list of fifteen symmetries, and in each case the relevant symmetry-operation is given, and the quantity which remains invariant to this operation. A third section shows whether the experiment has confirmed the validity of symmetry. We see that two of these symmetries are actually violated; in certain reactions, which are termed "weak interactions", strangeness and parity are not conserved.

For one who is not a physicist—and often for one who is—strangeness is such a strange concept that its non-conservation does not worry him. With parity it is another story. Has he not been assured that in elementary particle physics it corresponds exactly with what in the physics of our dimensions is called right-left symmetry of space?

Parity is something which simply cannot be visualized. Meaning, in its strictly verbal sense, the condition or state of being equal, parity in mathematics refers to the quality of being odd or even. Modern elementary particle physics uses the word in a quite narrow technical sense to designate a symmetry property of the wave function, that magic formula which in quantum mechanics describes the systems of particles which form our world. According to the object which it describes, this wave function, when all its space co-ordinates are reversed, either remains the same (symmetrical function, even parity) or changes sign (anti-symmetrical function, odd parity). Let us retain from this definition only the general impression of something discontinuous, of an either-or, of a brusque reversal—

actually all this is also characteristic of the right-left relationship —and let us trust the physicist when he tells us what the conservation of parity amounts to in practice: that a phenomenon and its mirror image are both equally possible; that every process occurring in nature can also occur as its mirror-image; that every "mirrored" movement is a movement permitted by the laws of nature—*in short that nature possesses right-left or mirror-symmetry*. Hence, whatever fate may befall its system, a plane of symmetry remains a plane of symmetry. Whatever happens to it and whatever transformations it may undergo, this equilibrium of possibilities remains, and what was in front of the mirror cannot get behind it.

When in 1924 Laporte—without actually using the expression parity—confirmed the mirror-invariance in the form of an empirical rule for atom transitions with emission of a photon, there was no hesitation in recognizing a basic quality of nature in this symmetry which within the macroscopic is accounted as trivial. It was bound, people believed, to dominate the steadily expanding science of nuclear physics as well. [42] Yet after the law of the conservation of parity had led to some notable discoveries (cf. p. 181), the ambiguity of certain nuclear processes—the so-called delta-theta puzzle—caused a number of physicists for the first time to entertain the idea, which at that time seemed rather absurd, that the law of the conservation of parity might have no validity in certain regions. From 1952 onwards Wick, Wightman and Wigner were reckoning with the possibility of discovering something of this kind but still considered it "fairly remote". [43]

It is to the Chinese physicists Lee and Yang, [44] working in the United States, that the credit is due for being the first to recognize the possibility of proving experimentally the conservation or non-conservation of parity and at the same time the irrelevance of all the previous attempts, which actually had no bearing on the question of violation of symmetry. They suggested certain experiments which were designed to provide information about the possible non-conservation of parity, and this work gained them the Nobel prize in Physics in 1957. Their analysis, which resembled Einstein's work in its concentration, was so profound that they would undoubtedly have earned this distinction even if the experiments which were subsequently made by the Chinese physicist Madame Wu and her collaborators had yielded a negative result. The result, however, was positive. It showed that parity, though conserved in processes with strong or electro-magnetic interactions, is not conserved in certain slow processes in which one of twenty or so weak interactions has a part to play.

The carrying out of the experiment was complicated by the necessity of having to use extremely low temperatures ($0 \cdot 001 \,^{\circ}\mathrm{K}$), but the basic idea behind it is really quite simple. Two experi-mental arrangements are set up, which are mirror-images of each other. Then an experiment is carried out, in which weak interactions are involved; it is carried out once with the right-hand arrangement and once with the left-hand arrangement, the object being to establish whether the same results (the same readings) would be produced. So in the first experiment of this kind nuclei of the radioactive cobalt isotopes (Co^{60}) were exposed to the influence of a magnetic field and the angular distribution of direction of electrons from the cobalt source was measured. Naturally it was part of the reversal that the electric current which generated the magnetic field should be sent through the coil once clockwise and once anti-clockwise.

"The whole experimental arrangement", says Wigner "has a symmetry plane and if the principle of sufficient cause is valid, the symmetry plane should remain valid throughout the further fate of this system. In other words, since the right and left sides of the plane had originally identical properties, there is no sufficient reason for any difference in their properties at a later time. Nevertheless, the intensity of the β radiation is larger on one side of the plane than the other side."[45]

This positive result was as surprising as the negative result of the Michelson-Morley experiment had been half a century before. An assymmetry of space, a distinction between left and right, seems to be made plain in this experiment in a most direct and obvious way. It is not "revealed" to us as the property of some impenetrable microworld, but we "perceive" it—as far as this term can be applied to particles—in our own large-scale world. The oddness of this result cannot be sufficiently stressed! Even for the naïve realist who "has always known what was right and what was left", but only intuitively, and who simply will not accept that they differ from one another in a measurable and communicable manner, the experiment really overshoots the mark. We are now really faced with the possibility of being able to telegraph to some "distant observer" a recipe which will enable him to distinguish right from left. "Put Cobalt nuclei at a temperature of $0 \cdot 001 \,^{\circ}K$ into a magnetic field and count by means of a counter the number of electrons emitted in each of the two opposite directions of the field. You will observe that they are emitted mainly in one direction. The side on which you can count the greatest number of electrons is the left and the other one is the right." But if left and right can be distinguished without a definition of direction, there must be an inner difference between the two, Leibniz must have been wrong in opposing Kant and Clarke, and the question they posed whether God could have made the right hand first and then the left must be epistemologically meaningful (cf. p. 148).

In reality it does not look as though the physicists have any such epistemological preoccupations. They assess the situation created by the non-conservation of purity much as they do that revealed by Mach's paradoxes. Here too the apparent asymmetry of space might reveal a hidden physical fact which had remained unobserved hitherto. Thus the physicist would be called upon to modify his description of reality in such a way that by taking this neglected fact into account *symmetry would again be restored*.

We can see what the nature of this neglected fact might be by looking at the two drawings illustrating the lecture which won the Nobel Prize for Yang and Lee (see appendix). On Lee's drawing the symbol of the cobalt isotope has been converted into its mirror-image. Does not this amusing detail suggest that the reversal of the experimental arrangement might not have been complete? The spatial reversal is plain to see. The magnetic field has also been reversed through reversing the electric current that created it. But how has the material used turned into its mirror image?

In our ordinary lives and in classical physics we never doubt for a moment that mirror reflection is a mere geometric-optical process and that the mirror reflects only our image without changing ourselves. But how is a micro-object, a micro-process reflected? *Is there perhaps such a thing as a "material" mirror image?* And is it perhaps essential for the complete reversal of the experimental arrangement that, where there is a weak interaction, *the matter must be replaced by the corresponding anti-matter?* The result of our experiment might then be explained by the difference between matter and anti-matter, a difference which may perhaps be no less significant than the difference between magnetism and electricity. We might well suspect that the experiment would not disclose any difference btween left and right if, instead of being carried out with Co^{60}, it was carried out with anti-Co^{60}. Unfortunately we know as little about anti-Co^{60} as we do about any other anti-atom. Anti-protons, positrons and other anti-particles react against one another, it is true, but they have never yet combined to form anti-nuclei or anti-atoms; moreover, they are rather difficult to work with, though physicists have recently succeeded in preserving them in bottles.

In science it sometimes happens, fortunately, that the sheer number of unsolved problems gives ground for hope that something known may ultimately emerge from several unknowns. So in our case it is an advantage that we are dealing not only with the invariance of the "parity operator" but also with the conservation of two other invariants: the *charge conjugation operator*, which changes every particle into its anti-particle, and the *time-reversal operator* which, as its name indicates, reverses time. This last law of conservation may need some explanation, as it runs counter to our idea that there must be an essential difference between the "known but uninfluenceable past and the unknown but influenceable future"—in other words that the arrow of time always points in the same direction. Reichenbach, Weyl and others have made it plain that the concepts of past and future refer to the *causal structure* of the world and that the concept of the unidirectional course of time enters physics only through the bias of statistical interpretation.[46] So far as the dimensions of nuclear physics are concerned, it is true to say that all laws are invariant to time reversal and that the film of events can be run off equally well in reverse. If a malignant spirit were to reverse time, the result would not be disorder but order in reverse. Everything would then happen as in August Kopisch's poem *Der grosse Krebs im Mohriner See*, in which the little poem finally flows back into the ink pot. The whole sequence of events would recur, but in reverse order. This has never yet been proved, but many physicists believe that the invariance of nature to time reversal could be proved experimentally, and experiments to this end are actually in progress at the present time. In any case the conservation of the time-reversal operator has never yet been seriously called in question.

Our three invariances of parity, charge conjugation and time reversal are not independent of each other. A very important theorem, the *CPT Theorem* (Charge, Parity, Time), first formulated by Lüders and Pauli, establishes a relation between them and so gives us a connection pattern for any violation of symmetry that may occur. This helps the physicist to escape from the dilemma caused by the result of the parity experiment. For supposing that the time operator is conserved he can conclude that the combined operation CP is valid: the right-left symmetry is conserved if all particles are transformed into their anti-particles. Hence our conjectures concerning the necessity for a "material" mirror-image is confirmed. Further, the connection expressed in the CPT theorem also makes it possible to substitute for an experiment that is incapable of being carried out—that with anti-matter—an experiment which it might well be possible to carry out, namely one concerned with the time reversal operator or the operation CP.

Only if some future Nobel Prize winner manages unexpectedly to prove that the invariance of the time-reversal operator is not absolute, shall we possess a method enabling us to distinguish right from left and even anti-matter from matter. Under our prevailing assumptions the CPT theorem teaches us that though parity as such may not be conserved, nor the charge conjugation which determines the character of matter or anti-matter, their *product*, i.e. their conjunction into a pair remains conserved. In the final analysis then, the discovery of the year 1957 has not led to the abandonment of right-left symmetry. Rather is it absorbed in a more perfect symmetry—such as not so long ago the symmetry of Newtonian space had to give place to the more perfect symmetry of Einstein's space-time continuum. The attempted fusion of two basic symmetries, that of space and that of matter, permits us to continue to regard the laws of nature as invariant to mirror images. Right and left continue to be indistinguishable in the Leibnizian sense, since the distinction between them depends on what we regard as matter and what as anti-matter. We may give our distant observer (cf. p. 177) the most minute directions on how he is to conduct his experiment and assure him to our heart's content that the side on which his counter gives the stronger reading is the left side and the other the right —all this is of no use to him if his matter is not our matter and his anti-matter is not our anti-matter.

Thanks to the ambiguity which he has discovered, the physicist is able to restore the symmetry of the whole which had seemed to be lost. In a bold vision that goes beyond anything yet imagined by science-fiction writers he can design the picture of a new hyper-symmetrical universe in which matter and anti-matter and right and left balance each other in the most intimate interconnection. Somewhere in the far distance, he guesses, there may be great quantities of anti-matter; half of the millions of galaxies could be anti-galaxies which resemble our own galaxies, like twin sisters—apart from the fact that they have the unpleasant characteristic of annihilating them and at the same time of being annihilated by them.

Meanwhile we do not know whether the world is really made like this and whether our means will ever permit us to find out. Little is known as yet about the behaviour of anti-matter. Is it "anti-gravitational"? If so this would explain its accumulation in distant regions. Does it, as Feynman believes, travel backwards in time? It is doubtful whether we shall ever be certain that we have observed a galaxy and an anti-galaxy locked in deadly embrace. It also seems questionable to regard the anti-world, which we may possibly have to imagine as permeating our own *hic et nünc*, as relegated to "the other corner of the universe". (But such localization is perhaps an unjustifiable visualization, and the strange images to which the new physics is leading us only appear strange because we see them "classically".) At all events, to nearly all physicists, and especially to Yang and Lee the restoration of symmetry through a switch to the anti-world looks like a triumph of scientific thought; it shows Yang that the structure of the symmetry laws conceals riches which we are still very far from understanding. For O. B. Klein[47] too, nature seems to be far richer than had ever been imagined, but also very different. We see here the new physicist's ideal, according to which the degree of perfection of a theory depends not on the simplicity of the structure it postulates, but on the breadth of the group to which the field-equations are invariant.

Einstein showed in the essay to which reference has already been made (cf. p. 156) how these demands get in each other's way, and pointed out that on grounds of phsyics the general theory of relativity sacrifices the simpler structure to the generality of the group. In solving the parity problem the physicist makes a corresponding choice. He wants nature to include mirror symmetry. That is why he demands invariance to the larger group, sacrificing the simpler structure to the generality of the group. Experience has shown him that the symmetry laws are only valid for a more complicated structure which brings matter into the spatial concept. One might well recall here a sentence written years before, and on another occasion, by Weyl, "Instead of assailing symmetry, we should search for another, richer structure, but one which again would have to be of the same nature everywhere". Do not these words prophetically foretell the development which led to the solution of the parity problem? Certainly it is no mere chance that Weyl's two-components theory of the neutrino has again become relevant and possible, thanks to this latest development in theoretical physics.

Radical though it may appear, the change in our basic principles resulting from the parity crisis, followed the general trend in the development of micro-physics. Any physicist who thought things through to their logical conclusion was bound to realize that the right-left symmetry with which we are familiar from our own dimensions cannot be transposed just as it is into the micro-world. The micro-world is not just a smaller edition of the macro-world. The fact that the concepts of space and time become inadequate here has already been recognized by Bohr when he discovered that the quantum transition transcended the space time frame (cf. p. 173). Hence the most startling thing in the whole parity crisis is perhaps the fact that God, as Pauli remarked, though He is left-handed in weak interactions, appears to be ambidextrous in the case of "strong" processes. The physicist is naturally concerned that the elementary processes should somehow lead to the laws of macroscopic physics, and that the conceptual framework which perception suggests should as de Broglie says emerge on the level of macrophysics through the *jeu des moyennes*.

From the epistemological point of view, the situation created by the non-conservation of parity only confirms what has long been known about the relation between the conceptual framework chosen for our description of nature and the observation of nature. "Reason", says Jean Ullmo in a noteworthy examination of modern thinking in physics, "no longer has a permanent content."[48] Though there are still *a prioris*, they are only hypothetical ones. The laws of conservation thus merely provide base points on which reason can rest and from which it continually takes off again. It is true that the objective world reveals itself to us in the groups but we do not know in advance in which. The only thing we can do is to try out certain groups, and the invariants which they provide, by bringing them into new contexts of experience. Sometimes they prove their worth. Sometimes we discover that they were not the right ones, that our supposedly *a priori* frame must be loosened, or that it was in fact only half a frame, part of a greater frame which we then regard as complete—though who knows for how long! Regarded in this perspective, symmetry, as a very elementary "rational demand", is in a very strong position when a sacrifice has to be made and we have to decide what shall be sacrificed. *Yet the maintenance of any particular symmetry cannot be the subject of any absolute claim on the part of science.*

Of course this does not mean that theories which lead us to such unexpected aspects of familiar phenomena as does right-left symmetry should be too readily accepted as self-evident.

I shall quote some critical observations by E. P. Wigner which will serve to remind us of that:

"The restoration of the correspondence between the natural symmetry properties of space-time on the one hand and the laws of nature on the other hand, is the appealing feature. . . . But maintaining the validity of symmetry planes forces us to a more artificial view of the concept of symmetry and of the invariance of the laws of physics. The other alarming feature of our new knowledge is that we have been misled for such a long time to believe in more symmetry elements than actually exist. There was ample reason for this and there was ample experimental evidence to believe that the mirror image of a possible event is again a possible event, with electrons being the mirror images of electrons and not of positrons. . . . We are now forced to believe that this symmetry in the experiments of Laporte is only approximate. . . . Further: there is the fact that molecules which have symmetry planes are optically inactive; there is the fact of symmetry planes in crystals. All these facts relate properties of right-handed matter to left-handed *matter*, not of right-handed matter to left-handed *anti-matter*. . . . This approximate validity

of laws of symmetry is, therefore, a very general phenomenon—
it may be *the* general phenomenon. We are reminded here of
Mach's axiom that the laws of nature depend on the physical
content of the universe, and the physical content of the universe
certainly shows no symmetry. This suggests—and this may also
be the spirit of the ideas of Yang and Lee—that all symmetry
properties are only approximate. . . ." Wigner declares himself
convinced that these discoveries "herald a revision of our
concept of invariance and possibly of other concepts which are
even more taken for granted".[49]

This same physicist, who prophetically divined the non-
conservation of parity in advance, cannot hide his discomfiture
over the fact "that the symmetry of the real world is smaller
than we had thought". Nevertheless, he seems in the end to
accept Mach's axiom and to be reconciled to the idea of living
in half a world, in a world that is incomplete in a more radical
sense than that expressed in Plato's myth. Actually I believe that
there is as little sense as there was four hundred years ago in
asking the question why the world is as it is and not otherwise,
or why God, as Kepler expressed it, has put the zodiac in one
place and not in another (cf. p. 144). A French physicist—
Astier—says quite rightly when speaking of the alignment of the
"spins", "I know very well that in this alignment the direction
itself is *a priori*. But the real world chooses. The world in which we live
is an a posteriori world."[50]

With this remark I might conclude my description of the
parity crisis if a question which has already been posed did not
once more present itself. If "our" real world were not entirely
right-left symmetrical, if that is to say, in our "world sector"—
that expression must not, of course, be too spatially interpreted—
if in our world sector, possibly because of an excess of matter
(or anti-matter!) right and left were displaced from their
proper relation to one another like heaven and earth in the
Chinese myth—could these special circumstances make life
possible and give it a special *direction*?

The thought is a tempting one, but the way from the elementary particle to the living world is long—longer than the way from mankind to the gods—to adapt a phrase from K. Reidemeister.[51] So the attempt to deduce an asymmetry of the organic from an asymmetry of our world sector must for the present seem as illusory as the attempt which certain physicists have made to deduce the spontaneity of biological happenings or even of freedom in the highest and noblest sense of the word from Heisenberg's uncertainty principle. I fear that this kind of interpretation of the findings of physics may prove disadvantageous not only to the scientist but also to the philosopher and, in the final analysis, to the religious man.

VII

CONCLUDING OBSERVATIONS:
THE MAGIC OF THE MIRROR

If ever a symbol deserved to be called an archetype—an organizing principle valid at different levels, it most certainly is the mirror. Even in our own day there is still a mystery of the mirror, though perhaps in a different sense from that discovered by the fortune-tellers and mirrored beauties in G. F. Hartlaub's *Zauber des Spiegels* (Magic of the Mirror). The mirror shows us the truth or conceals it; truth is then within it, and in this case the scientist, as the example of modern physics has shown, has the greatest difficulty in extracting the truth.

We have encountered the *illusion of the mirror*, in the most diverse forms, in connection with the right-left problem. As an archaic attribution of characteristic properties to the right or the left hand or the two corresponding directions in space, it led to the misleading concept of the psychological mirror image. In the field of biology it forced the infinite variety of the real into a deceptive pattern: twins were accounted either as identical or as enantiomorphically similar; their differences were disregarded, although these differences embodied the whole dialectic of heredity and environment. Finally, in physics, the concept of the mirror image, considered as a purely optical geometrical process, proved to be unsuitable for the reproduction of the whole physical content of certain symmetry operations. For the final solution of the parity riddle it was necessary, after the coupling of space with time by the theory of relativity, to attempt a coupling of space with matter. Hence the destruction of the spatial as an independent entity—the final stage—for the time being—of a process which began with the gradual annihilation of the perceptual world and has not even stopped at the conceptual frame of this world.

The mirror pattern prevents the human spirit from freeing itself from the spatial and fetters it to the rigid cross of the co-ordinates, We have every reason to believe that it is this determination to retain the spatial and to transfer it quite unjustifiably to unrelated fields and steps, which must be held responsible for all the paradoxes, discrepancies and blind alleys of thought that make their appearance whenever the concepts right and left are involved. In this respect the illusion of the mirror is only one facet, though a characteristic one, of that general "spatial cancer" which Bachelard recognized as a disease of human thought, though he knew nothing of the "de-spatializing" which had been brought about by the parity crisis. "Miserable place-adverbs" he says in his *Poétique de l'espace*,[1] "are given an insufficiently supervised ontological definition." Some of his observations concerning *outside* and *inside* could equally well be applied to *right* and *left*. "They produce a dialectic of dismemberment. This visible geometry makes us blind . . . it has the cutting sharpness of the yes-no dialectic which decides everything. Without being conscious of it, people make it the foundation for all ideas governing the positive and the negative. Thus the most profound metaphysic becomes rooted in an implicit geometry. . . . *Would the metaphysician think if he did not draw?*"

If he wishes to think without drawing, the metaphysician must guard against the "privileged evidences" which are peculiar to geometrical visualization and against all definitive intuitions. And the same applies to everyone who is willing to "follow the bold steps of the poets". Only theirs? The most recent development of physics has shown us forms of boldness to which only a de-spatialized thought can do justice.

Various answers have been given to the question whether in the new de-spatialized world which is gradually being opened to the human spirit anything will remain of the opposition between left and right. Perhaps their "being two" will be preserved—not with Pythagorean rigidity but in a more subtle and mobile dialectic—in the form of dual structures in which the sensible world and certain mathematical "entities" could both have a part. In his essay *Symétrie et Dissymétrie en Mathématiques et en Physique*, Albert Lautman[2] gives a number of examples of inwardly duplicate entities (so-called Spinors) and shows how a mixture of symmetry and dissymmetry is rooted in the very heart of algebra. By a vast detour which contains the whole history of human thought, such modern conceptions are undoubtedly bringing us back again towards the old myths of the *pair*.

APPENDIX

Figure 1
Carl Koch's tree test
(cf. p. 79)

Left emphasis   Right emphasis

Figure 2
Enantiomorphic crystals
(cf. note 4 to Chapter IV,
p. 195)

Figure 3a
The formation of twins
according to G. Dahlberg
(cf. pages 123 to 129)

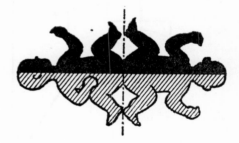

Figure 3b
Diagram of the formation
of twins according to
H. Newman (cf. pages
123 to 129)

$$Co^{60} \rightarrow Ni^{60} + \bar{e} + \nu$$

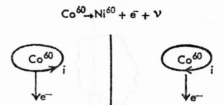

Figure 4a
The non-conservation of parity. Drawing by T. D. Lee in the Nobel Prize essay
(cf. pp. 175 et seq.)

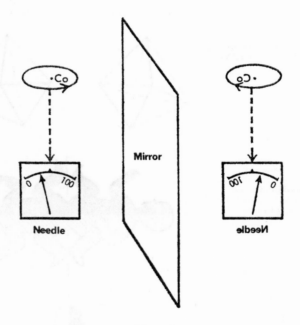

Figure 4b
The non-conservation of parity. Drawing by C. N. Yang

NOTES

NOTES

PREFACE
 1. Hermann Weyl, *Symmetry*, Princeton, 1953. The distinguished German mathematician and physicist has also made a valuable contribution to the philosophy of science. In this connection, see his contribution to the *Handbuch der Philosophie*, Section II, Munich–Berlin, 1927 and *Philosophy of Mathematics and Natural Science*, Princeton, 1949.

I. THE PAIR
 1. Gaston Bachelard (1884–1962), known for his "Psychoanalyses" of water, fire, etc., built up the whole of his thought upon the basic themes of the philosophy of science. His little book *Le nouvel esprit scientifique*, 1934, has lost none of its topicality. We shall make frequent reference to it, and particularly to the last chapter on the new non-Cartesian epistemology.
 2. Ibid.
 3. Henri Wallon (1879–1962), distinguished French psychologist and educationalist. His observations of wounded soldiers led him after the First World War to a genetic conception of psychology which differs from that of Piaget in essential points. In *Les Origines de la Pensée chez l'enfant*, Paris, 1947, he develops his theory of phases. In this connection see the *Handbuch der Psychologie*, vol. 3 (*Entwicklungspsychologie*).
 4. René Zazzo, *Les Jumeaux, le couple et la personne*, Paris, 1960.
 5. Cf. Erich Unger, *Wirklichkeit, Mythos und Erkenntnis*, Berlin, 1930.
 6. Robert Merle, *L'Homme, le rythme et la symétrie*, Paris, 1955.
 7. Mandala = a magic circle or square of eastern origin. For C. G. Jung symbol of the self. Considerable psychological importance therefore attaches to a "disturbed mandala". Cf. C. G. Jung, *Gestalten des Unbewussten*, Zurich, 1950.
 8. Marcel Granet, *La Pensée chinoise*, Paris, 1934. Cf. also *La Civilisation chinoise*, Paris, 1948.
 9. Joseph Needham, *Science and Civilization in China*, vol. II, History of Scientific Thought, Cambridge, 1956. The present little book owes much of its leading ideas to Needham's encyclopedic work.
10. Werner Müller, *Die Religionen der Waldindianer Nordamerikas*, 1956.
11. *Studium Generale*, vol. II, 1949.
12. Concerning romantic nature philosophy cf. the relevant chapter in K. E. Rothschuh, *Geschichte der Physiologie*, Berlin, 1953, with particular reference to Görres (1777–1848), *Principles of a New Foundation of Life through Dualism and Polarity*.
13. Arthur Schopenhauer, *Farbenlehre I*, Chap. 2, § 6. (Polarity of the retina and polarity in general.)
14. Schopenhauer, *The World as Will and Idea*, Book II, § 27.
15. Ibid., Book IV, § 56.
16. Schopenhauer I, Chap. 2, § 13.
17. Frey-Wyssling, *Submicroscopical Morphology of Protoplasm*, Amsterdam, 1953.
18. K. L. Wolf and W. Troll, *Goethes Morphologischer Auftrag*, 1942, consider this view "typically German". It has, they say, withdrawn into the background "since the spirit of western Enlightenment, which is inimical to morphological views, gained its alien dominion over the Germans".

II. IS RIGHT BETTER THAN LEFT?

1. Lucien Lévy-Bruhl (1857–1939), whose work on the mentality of the primitive found many an echo. In the *carnets* which were published after his death he dispenses with the concept of *participation*.
2. Mircea Eliade, *Das Heilige und das Profane*, 1951.
3. J. J. Bachofen (1815–1887). Despite the errors it contains, his work, *Das Mutterecht*, Bâle, 1861, opened new ways for ethnological research.
4. Richard Kobler, *Der Weg des Menschen vom Linkshänder zum Rechtshänder*, Vienna and Leipzig, 1932.
5. Ernst Siegrist, *Zur Händigkeit des Menschen*, Bâle, 1957.
6. Werner Speiser, *China, Spirit and Society*, English edition, 1960.
7. Joseph Needham, ibid.
8. Francis Warrain, *La Théodicée de la Cabbale*, Paris, 1949.
9. More details in Paul Althaus, *Der Schöpfungsgedanke bei Luther*, Munich, 1950.
10. A. Herschel, *Les Bâtisseurs du temps*, Paris, 1957. Cf. also Martin Buber, "Tales of the Hasidim", English edition, 1956.

III. THE EXPERIENCE OF RIGHT AND LEFT

1. Curt Elze, *Rechtslinksempfinden und Rechtslinksblindheit* in *Zeitschrift für angewandte Psychologie*, 24 (1924); *Kann Jedermann Rechts und Links unterscheiden?*, *Deutsche Zeitschrift für Nervenheilkunde*, 90 (1926).
2. It is not only hedgehogs that can be trained to know left and right. The same can be said of rats, guinea-pigs and even of earthworms.
3. We designate as kinaesthetic sensations the diffused feelings that a person has in regard to his own body.
4. Ernst Mach (1838–1916). The Austrian philosopher is accounted the principal representative of the older sort of positivism. He tends to be criticized today for completely mistaking the nature of the atoms, seeing nothing more in them than "symbols of those curious sensible elements that we encounter in the narrower fields of chemistry and physics". His *Analysis of Sensations*, English edition, 1959, is in fact outdated. As against this, the bold ideas developed in his *Science of Mechanics*, English Edition, 1960, approximate perceptibly to those of the theory of relativity.
5. Ernst Mach, *Populärwissenschaftliche Vorlesungen*, Leipzig, 1910, *Symmetry*—English edition, 1943.
6. Ibid., *Eine Betrachtung über Zeit und Raum*.
7. R. G. Natadze, *Die Hand als Faktor der Wahrnehmung der Raumrichtungen*, International Psychological Congress, Brussels, 1957. A discussion of these experiments by H. C. van der Meer (cf. note 18).
8. J. Kohler, *Umgewöhnung im Wahrnehmungsbereich*. This work, which was unfortunately not available to the author, describes according to H. C. van der Meer a repetition with improved equipment of the experiments by G. M. Stratten ("Vision without inversion of the retinal image", *Psychological Review*, 1897–4).
9. Schopenhauer, *The World as Will and Idea*, II, Chap. 4.
10. Homogeneity signifies the equivalence of all points in space, i.e. it means that *any* point can be brought into coincidence with *any* other through "motion". Isotropy signifies equivalence of directions. Straight lines passing through a point can be made to coincide with one another by a rotation around this point. Homogeneity and isotropy, along with three-dimensionality, form the characteristic properties of Euclidean space. Only in a homogeneous and isotropic space can the traditional concept of a rigid body be maintained.
11. Henri Poincaré (1854–1912). This important and versatile mathematician and physicist discovered amongst other things the automorphous functions. In his writings of a more general character, *Science and Hypothesis*—English edition, 1905, *The Value of Science*—English edition, 1958, *Science and Method*—English edition, 1956, *Dernières Pensées*, he practises the technique of the thought experiment with marked virtuosity. His conventionalism is often wrongly equated with the neo-positivist position.

12. Hermann von Helmholtz (1821–1894) one of the founders of thermodynamics, also enriched the physiology of sensations through his many discoveries. Medicine owes him the eye-mirror. His essay: *Über die Tatsachen die der Geometrie zu Grunde liegen* was written under the influence of Riemann's celebrated lecture: *Über die Hypothesen welche der Geometrie zu Grunde liegen* (1854).

13. According to Gottfried Martin, *Immanuel Kant, Ontologie und Wissenschaftslehre*, Cologne, 1915, Kant through his acquaintance with Lambert was familiar with the non-Euclidean geometries.

14. Jean Piaget, contemporary Swiss psychologist and logician. His great epistemological-psychological work, *l'Epistémologie génétique*, which he compiled in collaboration with his students from 1949 onwards (*La représentation de l'espace chez l'enfant, La Genèse du nombre chez l'enfant, La Géometrie spontanée de l'enfant, Le développement de la notion du temps chez l'enfant*, etc.) represents a unique "Embryology of reason". "In regard to the question whereon rests the gradual increase of achievement", says Wolfgang Metzger in *Handbuch der Psychologie*, vol. 3, "and above all what increase in cognitive functions and cognitive means lies at the bottom of it—to answer this question we have, save for the great series of researches by Piaget and his students, only scattered data." The reversible action or operation can be regarded as Piaget's basic concept.

15. "Vision should be looked on as an imperfect form of feeling, but one which takes place at a distance and uses light rays as long feeling rods" (*Über die vierfache Wurzel des Satzes vom zureichendem Grunde*, Chap. IV, para. 21.

16. *Albert Einstein, Philosopher-Scientist*, Evanston and Chicago, 1949. Ed. Dr Paul Arthur Schilp.

17. Jean Piaget, *The Psychology of Intelligence*, New York, 1950 (Harcourt Brace).

18. H. C. van der Meer, *Die Rechts-linkspolarisation des phänomeinalen Raums*, Groningen, 1958. The sentence quoted is in an essay by E. Straus.

19. Extraversion and Introversion are defined by C. G. Jung as inner directions of the libido which is either directed towards the outside and is in the service of action or is directed inwards and corresponds to a negative subject-object relationship.

20. Ludwig Klages, *Handschrift und Charakter*, 17–18th edition.

21. Max Pulver, *Symbolik der Handschrift*.

22. Charlotte Wolff, *The hand in psychological diagnosis*, London, 1951.

23. Karl Koch, *Der Baumtest*, Bern-Stuttgart, 1954.

24. Heinrich Wölfflin, *Gedanken zur Kunstgeschichte*.

25. Mercedes Gnaffron, *Die Radierungen Rembrandts*, Mayence, 1948.

IV. SYMMETRY AND DISSYMMETRY IN ORGANISMS

1. Goethe's Works, Weimar Edition Sec. II, 7, *Zur Morphologie*, Part II, 1892/93.
"Goethe's work on the spiral tendency of plants," says Troll, "came into being between 1827 and 1829 and was first published in 1831 as the last chapter of the Franco-German edition of the *Metamorphose der Pflanzen* (Metamorphosis of plants), later in detail in posthumous publications of his works."
In connection with the spiral tendency, cf. the concepts of polarity and ascending progress which Goethe describes in a letter to Chancellor Müller as "Nature's two great driving wheels" . . . "the one belonging to matter in so far as we conceive of it materially, the other belonging to it in so far as we conceive of it spiritually, the one being engaged in continual attraction and repulsion, the other in an unceasing effort to rise. But because matter can never exist or be effective without spirit nor spirit without matter, so matter too is able to undergo increase, so too it never abandons the desire to attract and repel; just as only that man can think who has separated enough to combine and has combined enough to wish to separate again."

2. Wilhelm Ludwig, *Das Rechts-Links-Problem im Tierreich und beim Menschen*, Berlin, 1932.

3. Louis Pasteur (1822–1895). The work in which the great chemist and biologist describes his discovery, bears the title "Memorandum of the relationship

which can subsist between crystal form and chemical composition and on the nature of rotation polarization" (22 May 1848).

4. Dissymmetry. This concept which was introduced by Pasteur contains all the ambiguity of the right-left problem; it is based on the occurrence of pairs of objects each of which is in itself asymmetrical, i.e. possesses no element of symmetry, and yet they stand to one another in the relation of image and mirror image like the little b and the little d. The crystallographer calls such crystals enantiomorphic.

5. Jacques Nicolle, *La Symétrie*, Paris, 1957
6. Ibid.
7. Goethe, ibid.
8. Wilhelm Ludwig, *Symmetrieforschung im Tierreich*, Studium Generale, vol. 2, 1949.
9. A. Portmann, *Die Tiergestalt*, Bâle, 1949.
10. Wilhelm Ludwig, ibid.
11. Hans Driesch (1867–1941). The chief representative of Vitalism rejects the "mosaic conception". "The prospective significance of every blastomere is a function of its position in the whole." (*Analytische Theorie der organischen Entwicklung*, 1894.)

V. THE LATERALITY OF MAN

1. Paul Broca (1824–1880). The starting point of his discovery was the discovery made in 1865 that injuries producing disturbances of speech frequently take place in the third frontal convolution of the left half of the brain.
2. O. L. Zangwill, *Cerebral Dominance and its Relation to Psychological Function*, Edinburgh, 1960.
3. Arnold Gesell. We are in debt to the American psychologist for numerous studies in developmental psychology. For the problem of laterality see especially A. Gesell and Louise B. Ames, "The Development of Handedness", *Journal of Genetic Psychology*, 1947, also the contribution in Carmichael's Handbook, 1946.
4. Magnus's tonic neck reflex occurs already in the 28-week-old embryo; when the head is turned, the arm and leg of the side in question are automatically stretched out, while on the opposite side the arm is bent at the elbow joint and the leg at the hip and knee joint.
5. Gertrude Hildreth, "The Development and Training of Hand Dominance", *Journal of Genetic Psychology*, 1949.
6. Richard Goldschmidt, *Understanding Heredity*, English edition, 1952.
7. Wilhelm Ludwig, *Das Rechts-Links-Problem*.
8. Newman, Freeman and Holzinger, *Twins*, 1937.
9. Rene Zazzo, ibid., vol. 1.
10. Ibid.
11. O. von Verschuer, *Genetik des Menschen, Lehrbuch der Humangenetik*, 1959.
12. Heinrich Bouterwek, *Rechts-Links-Abwandlung in Händigkeit und seelischer Artung*, *Ztsch. f. menschl. Vererbung und Konst*, 1938, 21.
13. Wilhelm Fliess, *Der Ablauf des Lebens*, Leipzig and Vienna, 1925.
14. The letters written by Fliess to Freud have been lost but the French Psychoanalyst Marie Bonaparte was able to save those written by Freud to Fliess. They were published in English as *Letters to Wilhelm Fliess by Sigmund Freud*, London, 1954. The remarks that are of special interest here occur in Letters 80 and 81.
15. F. Giese, *Psychologie der Arbeitshand*, 1928.
16. Even the leftward slope of the letters does not make writing easier for the left-handed. R. Zazzo (ibid.) examined the inclination of the writing of 244 twins and concluded that the inclination has nothing to do with handedness or at any rate "stands in a very indirect and complex relation to it".
17. Kenneth L. Martin, *Handedness, a Review of the Literature on the History, Development and Research of Laterality Preference*.

VI. THE WORLD OF THEORETICAL SCIENCE

1. Aristotle, *De Caelo*, Book II, Chap. 2, 4.
2. Galileo, Dial I op 1 quoted by Bruno Cassirer in *Das Erkenntnisproblem in der Philosophie der Wissenschaft der neueren Zeit*, vol. I, 1906.
3. Quoted by Cassirer, ibid.
4. Cf. note 2.
5. Max Jammer, *Conceptions of Space*, 1954, is the best work on space. Introduction by Albert Einstein.
6. Cf. Chap. III, note 10.
7. Immanuel Kant, *Vom ersten Grunde des Unterschiedes der Gegenden im Raum*, 1768. (On the First Ground of the Distinction of Regions in Space, in Kant's Inaugural Dissertation and Early Writings on Space, translated by John Handy, Chicago, 1928.)
8. Hermann Weyl, *Symmetry*.
9. First incomplete edition of the Leibniz-Clarke correspondence 1717. Modern English edition with notes: C. H. Alexander, *The Leibniz-Clarke Correspondence*, Manchester, 1956. The Clergyman Clarke adopted Newton's view in this last phase of the debate between the two great thinkers, a debate that had begun with the quarrel over priorities in regard to the differential calculus. Princess Caroline (Caroline Wilhelmine Charlotte of Anspach) not only forwarded Leibniz's letters but, like Prince Conti, took an active part in the correspondence.
10. Sir Isaac Newton, *Philosophiae Naturalis Principia Mathematica*, General Scholium after Definition VIII (the last of the definitions preceding Book I). A new edition of the English translation by Andrew Motte (1729), "Mathematical Principles of Natural Philosophy", appeared in Cambridge, 1934.
11. Leibniz-Clarke correspondence (cf. note 9), Leibniz's fourth letter.
12. Leibniz's letter to Conti (1715).
13. Leibniz's third letter.
14. Cf. note 5.
15. Weyl, *Symmetry*.
16. C. F. von Weizäcker, *World View of Physics*, English edition, 1952. A chapter of the book is dedicated to the "Idea of the Theodicy".
17. Gottfried Martin, *Der Begriff der Realität bei Leibniz*, Kantstudien, vol. 49, 1957/58.
18. Hans Reichenbach, *Die philosophische Bedeutung der Relativitätstheorie* in *Albert Einstein: Philosopher-Scientist*, Evanston and Chicago, 1949.
19. Aloys Wenzl, *Die Philosophischen Grenzfragen der modernen Naturwissenschaft*, Stuttgart, 1954, points out that Newton's theory had no need to wait for Einstein, nor even for the discovery of electro-magnetic phenomena to reveal its internal lack of harmony.
20. Lincoln Barnett, *Universe and Dr. Einstein*, London.
21. Henri Poincaré, *Science and Hypothesis*, New York, 1952.
22. The equivalence principle of the general theory of relativity is illustrated by the famous "experiment in thought" of the lift in space (cf. Aloys Wenzl and Lincoln Barnett; notes 19 and 20). We must note that this principle only has "local" validity (principle of the local equivalence of accelerated systems). For field theories in contrast to distant action theories such as the Newtonian, the concept of neighbourhood is characteristic. Every acceleration field can be changed into a gravitational field but every gravitational field cannot be changed into an accleration field.
23. Henri Poincaré, *Science and Hypothesis*.
24. Hermann Weyl, *50 Jahre Relativitätstheorie*, Die Naturwissenschaften, Issue 38, vol. 4, 1951.
25. Albert Einstein (cf. Chap. III, note 16).
26. Cf. Chap. III, note 13. As against this, for Heinrich Lange, *Über den Unterschied der Gegenden im Raum*, Kantstudien, vol. 50, 1958/59, the content of this essay is connected "so closely with the build-up of the transcendental aesthetic that it

would appear possible by refuting it to hit Kant's transcendental philosophy at a decisive spot". (Cf. Chap. VI, note 7.)

27. Kant in *On the First Ground of Distinction between the Regions in Space* (cf. Chap. VI, note 7).

28. Louis Couturat, *Les Principes des mathématiques*, Paris, 1905. Couturat along with Bertrand Russell is one of the rediscoverers of Leibniz as a precursor of modern logistics—H. Freudenthal, *Le Developpement de la notion de l'espace depuis Kant*, *Sciences*, Paris, 1959, points to three uses of the same sophism which in his opinion represents a "logical derailment on the part of Kant".

29. Kurt Reidemeister, *Raum und Zahl*, Berlin, 1957. The fourth chapter "Concerning the Differences between the regions in Space", deals with mathematical and ontological aspects of the right-left problem. According to Reidemeister the possibility exists in mathematics of recognizing "two objects as different, although the objects agree in their properties".

30. Immanuel Kant, ibid. Leibniz himself remarks concerning the Analysis situs (Math. VII, quoted from Cassirer, "I am not concerned to bring my *calculum situs* into proper form because till now we have only had *calculum magnitudinis* and so our analysis not *perfecta* but *ab elementis Geometriae dependens*. For me, however, the elements themselves must come out *per calculum* and in this respect I succeed quite properly. From this analysis all depends that is subject to *imaginationi distinctae*."

31. Hans Reichenbach in several writings. Compare especially *Die philosophische Bedeutung der Relativitätstheorie* (Chap. III, note 16).

32. Leopold Infeld, *Über die Struktur des Weltalls*, in *Albert Einstein*.

33. Paul Renaud, *Analogies entre les principes de Carnot, Mayer et Curie*, 1937.

34. Bachelard, *Le nouvel esprit scientifique*, 1934 (cf. Chap. I, note 1).

35. German in the French text.

36. Bachelard, ibid.

37. Bachelard, ibid.

38. A. Kastler, Pierre Curie, in *Les Lettres Françaises*, No. 759, 1959. Pierre Curie's essays on symmetry are to be found in his collected works, Paris, 1908, which have an introduction by Marie Curie. "The symmetry of phenomena", says Marie Curie, "was for him an intuitive concept. Incidentally few physicists have had such a knowledge of crystallographic forms and symmetry groups."

39. Bachelard, ibid.

40. According to Wigner (cf. note 43), the general theory of relativity is the first attempt to deduce a law of nature by selection of the simplest invariant equation.

41. Strangeness is a concept which has been introduced into nuclear physics in order to explain the unexpectedly slow disintegration of certain particles. The strange particles were discovered in 1947 by Rochester and Butler in the form of "V shaped events". For strangeness a conservation law is valid in the sphere of "strong" interactions. It expresses the experimentally confirmed fact that strange particles are always formed in pairs.

42. For nuclear physics (which started in 1933 with the discovery of the positive electron or positron) the concept of transformation is characteristic. In contrast to the atom the subatomic particle is not formed out of other smaller ones which have been joined together. Thus from a neutron there comes a proton, a positron and a neutrino, without these new particles taking up the same space as the former neutron.

43. Eugene P. Wigner, born 1902, received the Nobel prize for physics in 1963 for theoretical work in the field of nuclear physics, work that is closely connected with the problems of symmetry. Making this their starting point the other two Nobel prizewinners of 1963, Maria Goeppert and Hans Jensen, reached a new conception of the atomic nucleus and developed a nucleospectroscopy. Forty years previously Wigner, as C. N. Yang pointed out in his Nobel Prize lecture, had taken a decisive step by recognizing the empirically discovered rule of

Laporte as a "consequence of the reflection invariance or right-left symmetry of the electro-magnetic forces within the atom".

44. Chen-Ning-Yang (born 1922) since 1949 Professor at the Institute for Advanced Studies in Princeton. Tsung-Dao Lee (born 1926) since 1951 Professor at Columbia University in New York. In what follows here the references are to the Nobel prize lectures: C. N. Yang, *The Law of Parity Conservation and other Symmetry Laws of Physics* and T. D. Lee, *Weak interactions and Non-Conservation of Parity* in *The Nobel Prizes, 1957*, Stockholm, 1958.

45. Wigner in *Reviews of Modern Physics*, 1957.

46. The time problem according to H. Reichenbach is concerned not with the order but with the direction of time. "Sooner" and "later" are—contrary to "before" and "after"—structurally different. "We cannot intend to go to the theatre yesterday" (*Les Fondements logique de la mecanique des quanta*).

47. O. B. Klein, "The 1957 Nobel Prize for Physics", in *The Nobel Prizes 1957*, Stockholm, 1958.

48. Jean Ullmo, *La Pensée scientifique moderne*, Paris, 1958.

49. E. Wigner, ibid.

50. A. Astier, *De l'élementarité des particules ou la dialectique de la substance, Revue de l'Association des Anciens Elèves de l'Ecole Polytechnique*, 1962.

51. "The way from men to the gods is much nearer than the way from men to this real body built up out of atoms." R. Reidemeister, *Geist und Wirklichkeit*, 1953.

VII. CONCLUDING OBSERVATIONS: THE MAGIC OF THE MIRROR

1. Gaston Bachelard, *La Poétique de l'espace*, Paris, 1957.

2. Albert Lautman (1908–1944). The French mathematician, who was shot in the concentration camp of Souges, only left a small number of essays with noteworthy contents (in addition to *Symétrie et Dissymétrie en Mathématiques et en Physique*, Paris, 1946, to which reference has been made, and which was published by Mme Suzanne Lautman, cf. *Les Schémas de Génèse et les Schémas de Structure*, Paris, 1938). In these essays a form of Platonism finds expression which is adapted to modern science. Mathematics are more than a mere language; there is a mathematical universe, filled with "mathematical beings". In particular, Lautmann draws attention to anti-symmetrical beings such as spinors, characterized by "an inner duality, which can be distinguished within one and the same being". Lautmann comes close to many neo-Kantians, including Brouwer, of whom Weyl in the *Handbuch* (cf. note 1 to Preface), says that like Plato he sees the root of mathematical thought in the "unity of two".

INDEX OF NAMES